DEPARTURES

TIME	FLIGHT	DESTINATION	GATE	REMARKS
11:15	UA411	NEW YORK	A17	CANCELLED
11:20	LH417	FRANKFURT	A26	CANCELLED
11:30	AF204	PARIS CDG	C03	CANCELLED
11:45	BA616	LONDON	B09	CANCELLED
11:50	AC702	TORONTO	A16	CANCELLED
12:00	CA101	BEIJING	B41	CANCELLED
12:15	JJ333	SAO PAULO	C36	CANCELLED
12:25	JL226	TOKYO	A07	CANCELLED
12:45	QF112	SINGAPORE	C11	CANCELLED
12:55	AZ651	ROME	B39	CANCELLED
13:10	AZ651	MOSCOW	A02	CANCELLED
13:20	QF204	SYDNEY	C08	CANCELLED
13:30	UA512	CHICAGO	A12	CANCELLED

NAVIGATING TURBULENCE

An Analysis of Airline Failures 2019-2023

ADIEL MAMBARA

BlueRose ONE.com
Stories Matter
NewDelhi • London

BLUEROSE PUBLISHERS
India | U.K.

BlueRose ONE.com
Stories Matter
NewDelhi • London

For permissions requests or inquiries regarding this publication,
please contact:

BLUEROSE PUBLISHERS
www.BlueRoseONE.com
info@bluerosepublishers.com
+91 8882 898 898
+4407342408967

ISBN: 978-93-6261-070-6

Cover design: Shivam
Typesetting: Namrata Saini

First Edition: July 2024

Author's Biography

Mr. Adiel Mambara is a highly accomplished executive in the aviation industry, with a distinguished career spanning over 28 years. For more than 25 years, he has been a prominent figure in the airline sector, where he has demonstrated his exceptional business acumen and leadership skills.

Adiel's current role as the Country Manager for Royal Brunei Airlines in the United Kingdom (UK) is a testament to his ongoing commitment and relevance in the industry. He oversees the airline's regional commercial operations in this position, ensuring its continued success.

Adiel's involvement in various industry and academic associations, such as the Board of Airline Representatives in the UK, Foreign Airline Association UK, and the University of West London Alumni Association, clearly indicates his commitment to the industry's growth. He actively contributes his expertise and experiences to further the interests of the industry and the academic community.

His academic credentials are equally impressive, having earned a Bachelor of Arts (BA) Honours in Business Studies with Marketing from the University of West London. He was also the first graduate to obtain a Master of Science (MSc) in Travel Business Leadership from Leeds Beckett University. The Institute of Travel and Tourism and Zimbabwe Achievers Awards has recognised his success and contributions to the industry.

Apart from his professional and academic achievements, Adiel also mentors undergraduate and postgraduate students at various universities. He draws from his extensive experience in the industry to offer guidance on educational and career-related matters in business, aviation, and travel and tourism.

Finally, he is no stranger to writing and has significantly contributed to various aviation-related subjects in magazines and newspapers. His expertise and insights on the industry are highly valued and sought-after by professionals and enthusiasts alike.

Acknowledgement

I am humbled and grateful to express my profound appreciation to the numerous individuals who have directly and indirectly contributed to the creation of this book. It has been an incredible journey filled with ups and downs, but the constant support and motivation I received from those around me kept me going.

I have had the great fortune of meeting people who have positively impacted my life in countless ways. Their kindness, generosity, and wisdom have been invaluable to me throughout the writing process. I am indebted to them for their unwavering support and encouragement.

I want to give special recognition to one individual, my unsung hero, who went above and beyond in proof-reading the entire book and providing invaluable feedback. Their attention to detail and constructive criticism have undoubtedly made this book better than it otherwise would have been.

Last but not least, I am deeply grateful to my family for their constant love and support that has been a guiding light on my journey.

This book is dedicated to:

"The memory of my beloved father, Noah Adiel Shava Mambara, who now resides among the stars. I hope this book find its way to you."

"My mother, Mrs. Tendie Mambara, who instilled in me the gift of dreaming and the courage to pursue it relentlessly."

“To those rare, exceptional relationships that have the power to for forever transform our lives.”

“To all those who have taken the time to read this book. Your willingness to give my words a chance is truly appreciated and I hope that you find this book to be informative and thought-provoking”.

Acknowledgement

Preface

Welcome to the first edition of Navigating Turbulence: An Analysis of Airline Failures 2019-2023

The book contains a comprehensive compilation of case studies that focus solely on the failures of various airlines, some of which have even gone bankrupt. The author has critically analysed these failures by examining the relevant background context, identifying the reasons behind these failures, and conducting a thorough analysis of each case study.

The book aims to give readers a deeper understanding of the airline industry's complex challenges. Each case study in the book follows a specific format and includes notable features.

The author presents a highly insightful and critical analysis of several airlines' business strategies and practices that have faced bankruptcy, failure, and eventual closure.

Through an extensive examination of various case studies, the book provides an in-depth understanding of the factors that led to the downfall of these airlines, including their operational inefficiencies, financial mismanagement, and inability to adapt to changing market dynamics.

The author's comprehensive and objective evaluation of these failures makes this book a valuable resource for anyone interested in the aviation industry and its challenges.

The format of each case study is structured as follows:

- Each case study includes an abstract at the beginning that provides a comprehensive yet concise summary of the essential technical information. The abstract is written in clear and easily understandable language for readers.

- The background section of each case study serves as a roadmap for the reader, providing a comprehensive understanding of the airline's history. The background offers a brief overview of the events that led to the airline's failure, emphasising the crucial points. This gives the reader an insight into what to expect from the rest of the case study.

- Industry analysts and senior airline managers have offered numerous explanations for why airlines go out of business. In this section (reasons for failure), we'll explore some of the common reasons for failure, such as poor management, weak financial foundations, and brand image conflicts resulting from changes in strategy or product evolution. While stakeholders may have unique perspectives on each case, similar factors are often at play in most instances. For instance, growth-related issues are a recurring problem.

- In order to provide a comprehensive and informative analysis, each case study will feature a concluding section that effectively summarises the main points and key takeaways from the analysis of the airline failure. This conclusion will serve as a valuable reference point for readers, helping them to better understand the factors that contributed to the airline's collapse and the lessons that can be learned from the experience. By distilling complex information into a concise and accessible format, the conclusion will help ensure that readers understand the issues from the case studies.

In the book's concluding chapter, the author comprehensively summarises all the major points covered. This chapter aims to help readers remember the key takeaways from the book and provide a sense of closure to the overall narrative. The author has also included his reflections gained while writing the book. Overall, the final chapter is an essential component of the book that allows readers to absorb and appreciate the knowledge and information presented fully.

Note from the Author

As a seasoned airline professional with over 25 years of experience, I have always been fascinated by the aviation industry and its inner workings.

The COVID-19 pandemic brought unprecedented challenges to the airline industry worldwide. It has particularly piqued my interest in the airlines that failed before and after the pandemic.

In response, I have written this book to provide valuable resources for anyone who shares my passion for aviation and wants to understand the industry's challenges.

My book comprehensively analyses the factors that cause airline failures, covering both macro and micro-level considerations. It delves into the reasons that have led airlines to go out of business, examining the economic, regulatory, and operational challenges that airlines face, and how these factors ultimately led to their failure.

Moreover, the book provides a nuanced understanding of the strategies employed by airlines to mitigate these challenges and ensure long-term survival.

By understanding the underlying causes of airline failures, we can develop effective strategies to ensure the industry's resilience and sustainability.

I want to express my gratitude to my readers for choosing to purchase my first book. With your support, I am confident I can write more books that will continue illuminating the aviation industry's intricate workings.

Contents

The airline business is the biggest team sport in the world. When you're all consumed with fighting among yourselves, your opponents can run over you every day.

-Gordon Bethune

CHAPTER ONE

An Overview of the Airline Industry – A Global Overview

Abstract

The airline industry has been experiencing significant volatility cycles in recent years, primarily due to frequent exogenous shocks. The COVID-19 pandemic was one such shock that brought the global economy to a standstill and necessitated economists to modify their established forecasting models to predict growth patterns in a post-pandemic world.

Since the pandemic-related travel restrictions were lifted in 2022, the air travel market recovery has been asynchronous and unstable at times, but the overall trajectory is positive. According to the International Air Transport Association (IATA), the airline industry was 2.5% below 2019 levels in December 2023.

Despite this, the strong performance in Quarter 4 of 2023 has given airlines hope for a return to standard growth patterns in 2024.

Chapter One of this book provides a comprehensive overview of the airlines' current state as we approach the end of 2023 and the start of 2024. It delves into the industry's challenges and opportunities.

Keywords: Airline, Pandemic, Profitability, Traffic

Current Issues facing the aviation industry worldwide

The global economic growth rate is expected to be around 3% in 2023, a slight decrease from the growth rate of 3.4% in 2022 and 6.3% in 2021. This is consistent with the average global GDP growth rate since the 1970s. While the airline industry has experienced a significant increase in air travel and a boost in profits following the pandemic, the industry's profit margins remain lower than pre-pandemic levels.

IATA predicts the industry's profits will reach over $23 billion in 2024. However, this figure is relatively small compared to the aggregate across the industry. Despite the challenges faced by the industry, airlines remain an attractive prospect for many investors.

Many new passenger airlines have emerged both before and after the pandemic. This trend of adding new start-up airlines to the market is common following a downturn. However, the industry is not without risks.

Several factors, including the economic and geopolitical environment, could threaten the airline industry's outlook. Specifically, geopolitical uncertainty is high in 2024, called the 'Year of Elections'.

Several significant elections are scheduled to take place in countries such as Indonesia, Russia, India, Mexico, the European Union (EU), the United States of America (USA), and the United Kingdom (UK). These countries account for over 50% of Gross Domestic Product (GDP) according to IATA's January 2024 key risks report (See Figure 1). The elections could lead to trade impediments, increased oil price volatility, and possibly more airline failures.

Figure (Fig) 1: 2024 Country Elections

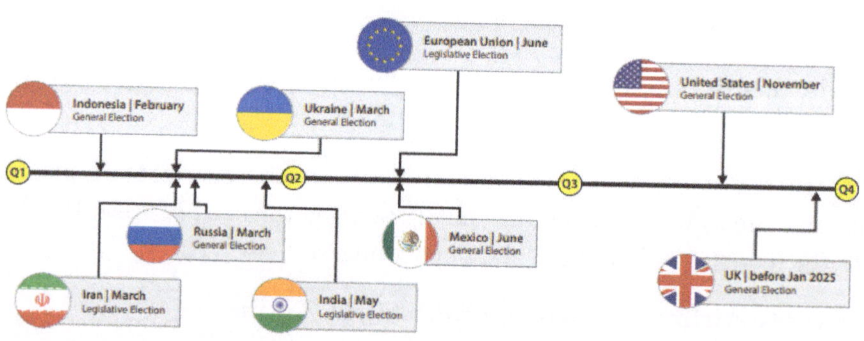

Source: BOC Aviation

The pandemic of 2020 significantly impacted the airline industry, resulting in a collective loss of $140 billion, according to a report by IATA. While airline failures have been relatively low, the recovery process has not been without its challenges, given historical norms.

The financial losses have forced airlines to look for new ways to stay afloat. Some entrepreneurs have taken advantage of the gaps in the market created by the demise of certain airlines and the apparent weakness of surviving airlines weakened by the pandemic and constrained by heavy debt burdens.

In addition to the pandemic's impact, airlines have been directly affected by supply chain issues that continue to impact global trade and business. Aircraft parts supply chain ruptures have negatively impacted the delivery of new aircraft and the ability of airlines to maintain and deploy existing fleets. This has been a challenge for aircraft and engine manufacturers, who have been unable to address the issue to date fully. According to the Aviation Leaders Report 2023, this issue has significantly impacted the airline industry.

Furthermore, the war in Ukraine and Gaza has had a significant impact on profitability for most airlines. Any potential escalation could have devastating effects. Between such scenarios, a fraught geopolitical environment is weighing on international trade and skewing risks to the airline industry's outlook to the downside, as per the IATA report of 2023.

The Middle East delivered a solid financial performance in 2022, likely recording a net profit of around USD 1.4 billion, with a 2.6% margin. The Middle East carriers have been swiftly rebuilding their international networks, and the region's financial recovery was supported by a significant increase in the passenger load factor of almost 25 percentage points in 2022, outstripping the performance of the other areas.

According to the IATA report of 2023, the net profit is expected to be around USD 2.0 billion at a 3.8% margin in 2023. On the other hand, African airlines are expected to generate a moderate loss of around USD 484 million in 2023. Africa remains a challenging market in which to operate an airline, with economic, infrastructure, and connectivity challenges impacting the industry's performance.

Conclusion

The airline industry has faced unprecedented challenges over the years, especially during the COVID-19 pandemic (declared by the World Health Organisation (WHO) as officially starting on the 11th of March 2020 and ending on the 5th of May 2023). However, despite the hardships, airlines have shown remarkable resilience, the hallmark of their response to the pandemic.

The recovery in 2021 was driven primarily by large domestic markets, which helped to offset some of the losses incurred during the pandemic. Throughout 2022, international traffic substantially recovered across the intra-European, Americas, and Transatlantic markets, which are now predominantly back to or close to 2019 levels. As we look towards 2023, it's expected that the industry will experience a financial recovery, marking the first industry profit since 2019.

This growth is expected to continue in the coming years. One key factor contributing to this recovery is the evident pent-up demand for travel, accompanied by an increased appetite for premium leisure travel. This demand has compensated for any reduction in business travel, which has been slower to recover. Given the significant global GDP increase since 2019, much growth remains recaptured, which bodes well for the industry's future.

In the next chapter, we will examine an overview of airline failures and bankruptcy

CHAPTER TWO

Overview of Airline Failures and Bankruptcy

Abstract

The airline industry has been extensively studied from an economic standpoint, but research focusing on the airline executives themselves, rather than just the airlines they manage, has been relatively scarce. Despite emphasising objective financial and operational measures to determine success, airline strategists subjective notions of success and failure have not been thoroughly explored.

This book seeks to delve into these notions and identify the factors that such strategists perceive as contributing to failure in different airline case studies presented in this book. The most apparent manifestation of airline failure is bankruptcy, and this book aims to uncover the underlying causes of such failures.

Key Words: Airline, Bankruptcy, Financial, Strategies.

Discussion

The airline industry is known for its high rate of failures, which has led to an old adage that 'the best way to become a millionaire is to start an airline as a billionaire'. However the definition of success and failure in the airline industry is not straightforward, and it depends on the perspectives of those involved in decision-making.

On one hand, the subjective notions of success and failure are held by those directly involved in determining an organisation's strategy. These notions are personal and subjective and are best investigated through a phenomenological approach.

On the other hand, objective performance-based notions of success and failure consider the financial and operational performance of airlines and are best investigated through a positivist approach that compares results with objective benchmarks.

The dichotomy between the positivist and phenomenological approaches is evident in the airline industry literature reviews. Academics have primarily researched airline failures from an economic perspective, but have often overlooked the insights of airline executives who are better positioned to explain how and why strategies are chosen.

Furthermore, the limitations of the economic perspective are significant, as the objective reality of the researcher may not be relevant to strategists, who may have different priorities and decision-making processes. Thus, understanding the factors that contribute to airline success or failure requires a more nuanced and comprehensive approach that takes into account both objective and subjective perspectives.

The aviation industry is irreplaceable in the global transportation system and is vital for many countries economic prosperity and strategic development. Not only has it demonstrated long-term resilience, but it has also weathered crises, proving to be an indispensable means of transport.

The aviation industry has recently been grappling with a surge in airline insolvencies and bankruptcies. The impact of these events has been quite significant, causing disruptions in market function and competition.

Bankruptcy or insolvency is an economic state where individuals, enterprises, or even public institutions accumulate debts they cannot repay. This can have far-reaching consequences, and governments often respond with various legal and policy measures to mitigate the situation. As stated by (Bellovary et.al.,2007), these responses can take various forms, including financial assistance, regulatory interventions, and market restructuring.

These responses under civil law can typically be grouped into two broad forms: liquidation and reorganisation. Liquidation involves the sale of an enterprise's remaining assets, which are then apportioned among its creditors. On the other hand, reorganisation involves creditors agreeing to become owners of a reorganised company, accepting equity in exchange for giving debt (Bellovary et.al.,2007).

In reorganisation cases, the company may continue providing services during the period it is being reorganised. While bankruptcy can be a powerful tool for airlines to turn their fortunes around, it is essential to note that it is a clear sign of financial distress caused by poor management or operational inefficiency. However, leading carriers like Delta, American Airlines, Air Canada, and United Airlines have utilised bankruptcy protection to facilitate significant restructuring or labour union negotiations, ultimately emerging stronger (Iverson,2018).

In Europe, several airlines have been forced to file for bankruptcy or cease operations over the last few years, with a shocking 301 airline failures recorded in this decade alone. The number continues to rise, amounting to more than two airline failures monthly. Europe simultaneously witnessed the highest number of start-ups, and the most failures (Dunn,2017).

As evidence of the industry's dynamic nature and evolving market conditions, **Table 1** lists the airline bankruptcies and restructuring in 2023, while **Table 2** shows the new airlines formed in the same year.

Table 1: Airlines Bankruptcies/Restructurings in 2023

NUMBER	AIRLINE	COUNTRY
1.	Cascadia Air	Canada
2.	Flybe	UK
3.	Flyr	Norway
4.	Novair	Sweden
5.	Aeromar	Mexico
6.	Viva Air	Columbia
7.	Ultra Air	Columbia
8.	Niceair	Iceland
9.	GoFirst	India
10.	Fly Gangwon	South Korea
11.	Air Moldova	Moldova
12.	Thai Smile	Thailand
13.	Buta Airways	Azerbaijan
14.	Swoop	Canada
15.	MY Airways	Malaysia

16.	JC International Airlines	Cambodia
17.	Equair	Ecuador
18.	Hi Air	South Korea

Source: KPMG Aviation Leaders Report, 2023

Table 2: New Airlines in 2023

NUMBER	AIRLINE	COUNTRY
1.	Ghana Airlines	Ghana
2.	Aerus	Mexico
3.	Centrum	Uzbekistan
4.	Air Samarkand	Uzbekistan
5.	Sky Vision Airlines	Egypt
6.	Air Moana	French Polynesia
7.	ESAV Airlines	Ecuador
8.	JetSMART Columbia	Columbia
9.	City Airlines	Germany
10.	Silk Avia	Uzbekistan
11.	Air 1	Iran
12.	Global Airlines	UK
13.	Mexicana (2.0)	Mexico
14.	Discovery	Germany
15.	Valetta Airlines	Malta
16.	AJet	Turkey
17.	BeOnd	Maldives
18.	NG Eagle	Nigeria
19.	Naysa	Canary Islands Spain
20.	Bermudair	Bermuda

Source: KPMG Aviation Leaders Report, 2023

Unfortunately, even well-managed airlines have been forced to file for bankruptcy due to uncontrollable events, such as the COVID-19 pandemic, which fatally disrupted their liquidity. It is therefore, crucial for airlines to take proactive measures to mitigate potential financial distress and ensure long-term sustainability, and survival.

Survival is a crucial strategic objective that must be prioritised by organisations operating in highly competitive, and volatile sectors, such as airlines. However, it is often ignored, leading to bankruptcies and failures, which have become increasingly common in the air transport sector, causing spillover effects on the public and financial distress.

Financial distress predictions are critical in providing early warning signals about future economic health to airlines and their stakeholders, including investors and regulators.

The airline industry has always been under immense pressure due to the extreme impact that financial distress can have on its balance sheet. Despite research limitations, it is evident that the underlying financial instability has persisted throughout the airline value chains.

CHAPTER THREE

The Collapse of One of Europe's Up and Coming, Low-cost Carrier – Romanian Blue Air

Abstract

The aviation industry in Romania has been active for nearly a century since the first experiments were conducted shortly before the First World War. Thanks to low-cost carriers, air travel has expanded rapidly in recent years, growing faster than the country's overall economic growth.

In the early 2000s, Romania was considered one of Europe's most promising emerging markets, with its youthful population and rapidly expanding middle class helping the aviation industry flourish. In 2004, Blue Air, a new airline, capitalised on this exponential growth. Within four years, the airline had become Romania's leading carrier, carrying over 1.6 million passengers.

One of Blue Air's unique strengths was ensuring the mobility of many citizens working abroad. Additionally, local entrepreneurs could travel between the various regions of Romania and to nearby European countries, which improved regional connectivity and contributed to the economic life of those regions.

Key Words: Air Transport, Bankruptcy, Debt, Growth.

Background

Blue Air was an airline that started operations in Romania in December 2004 with only two Boeing 737s. It was initially owned by Nelu Lordache and offered low-cost return flights from Băneasa to Barcelona, Lyon, and Milan-Bergamo for just €19. Despite gaining notoriety for delayed flights, especially among passengers flying to Barcelona, Blue Air eventually established itself as a reliable airline in the aviation industry (Blue Air, 2023).

In 2006, Blue Air became Romania's leading low-cost airline, transporting 0.4 million passengers and operating 50 weekly flights during the summer. Despite fierce competition from Wizz Air, Germanwings, and Easy Jet, the airline transported a remarkable 1.6 million passengers in 2008 and became Romania's largest domestic airline (Blue Air, 2023).

However, Blue Air's success was short-lived, and the airline ran into financial difficulties in 2009, forcing it to suspend operations. Although it resumed operations in 2010, it struggled to regain its former glory.

Despite its importance to the country's aviation industry, Blue Air struggled to recover after the global financial crisis in 2009. By 2012, it became clear that the airline was in severe financial trouble.

The Romanian government refused to bail the airline out, and Blue Air had no choice but to file for bankruptcy in 2013. After a two-year hiatus, Blue Air restarted operations in 2015 with a hefty growth plan.

In 2017, the airline witnessed a substantial upsurge in passenger traffic, transporting over 5 million passengers, signifying a remarkable 40% rise from the previous year's 3.6 million passengers. This notable feat cemented Blue Air's status as one of Romania's most thriving airlines.

According to Euro control records, as of September 2022, Blue Air provided services to 75 scheduled destinations across 21 countries.

Unfortunately the airline racked up too much debt, and eventually in 2021, the Court of Bucharest granted the request to initiate general insolvency proceedings for Blue Air, and the airline ceased operations in 2022

The reasons for the airline's downfall remain unclear, but specific issues which will be discussed below might have played a role (Journal, 2023).

Why Did Blue Air Fail?

The impact of Covid-19 and financial pressures

The spread of the coronavirus (COVID-19) severely impacted the global economy, leading to countries implementing measures to halt the pandemic.

These measures included social distancing, travel restrictions, and lockdowns, significantly affecting economic activities, particularly tourism and transport. Like many other transport companies, Blue Air suffered a devastating blow from the pandemic.

Air traffic fell by over 90% after restrictions aimed at limiting the virus spread, leading to a drop in receipts by over € 100 million compared to the level estimated and planned in March-June 2020.

The company stopped regular commercial flights, and over 90% of its employees were technically unemployed. Blue Air also downsized its fleet from 32 aircraft in 2019 to only 13 (European Commission, 2023).

In July 2020, Blue Air announced its entry into a pre-insolvency procedure, a form of restructuring to reach a formal agreement with its creditors regarding the payment of debts to ensure the company's continuity and viability (Dunn, 2023).

To mitigate the impact of the pandemic, Blue Air requested a rescue loan from the state. The loan, a preferential interest loan given to companies in significant crises, was not a non-repayable fund. The company received RON 300.7 million (approximately 51 million British pounds) as a loan with state guarantees to compensate for the effects of the COVID-19 pandemic. While the European Commission approved the loan, the letter of guarantee warned of the significant risk of execution and non-recovery of the amounts paid by the state (European Commission, 2023).

Blue Air also received state aid of (€28 million) in the name of public guarantees and an additional €34 million as a rescue loan to cover the company's liquidity for six months. The EU temporarily eased state aid rules in August 2020 in light of the pandemic's slowing down, but the European Commission investigated suspicions that the Romanian Government was supporting grounded local carrier Blue Air by breaking state aid rules (European Commission, 2023).

At the beginning of September 2022, Blue Air announced the suspension of all flights operated by the airline in Romania. The suspension was compounded by the Romanian Ministry of the Environment blocking Blue Air's accounts for non-payment of a fine, rendering the airline incapable of serving its customers and paying for essential flight services such as airports, handling companies, and fuel suppliers. Additionally, Blue Air was in enforcement proceedings for a fine representing the value of greenhouse gas emission certificates related to carbon dioxide emissions generated in 2021, which are paid by all airlines (Journal,2023).

The European Commission had classified Blue Air as a company in difficulty on December 31, 2019, before the pandemic. It could be argued that Blue Air was already experiencing financial difficulties before the pandemic.

Moreover, the airline was been losing money due to extensive investments made since 2016 to improve its network of routes. While the airline returned to profitability in 2019 and early 2020, the pandemic accelerated the airline's losses, resulting in urgent liquidity needs, just like other firms in the aviation sector.

Mounting Financial Penalties

Blue Air faced significant challenges that resulted in severe disruptions to its services. As a consequence of its actions, the National Authority for Consumer Protection (ANPC) imposed a hefty fine in July 2022 of over €2 million for Blue Air cancelling a staggering 11,289 flights between April 2021 and April 2022, affecting as many as 178,405 reservations (European Commission,2023). The fine was the largest ever handed down by the consumer protection authority.

The airline was targeted after the ANPC received more than 1,000 cross-border complaints. While this situation was unfortunate, the airline tried working with the ANPC to ensure that all regulations were followed (European Commission,2023).

The ANPC further accused the airline of accepting payments from promotional campaigns but then cancelling the flights, thus breaching EC Regulation 261/2004. In light of this, Blue Air was instructed to reimburse its customers within ten days of any cancellation.

The imposed fine pertained to all cancelled tickets during the third, fourth, and fifth pandemic waves. In addition, Blue Air voiced its discontent over the absence of similar obligations imposed on other companies engaging in comparable practices. (Journal, 2023).

Fleet Issues and Flawed Business Model

Blue Air's business model shifted to a low-cost air carrier in 2020, which made it impossible for them to generate revenue from charter flights or leasing aircraft with crew, maintenance, and insurance included. In 2019, they were able to do so, but not anymore.

Even though Blue Air has used six different aircraft versions from Boeing's 737 family, including all three versions of the 737 Classic series (-300, -400, and -500 models), the average age of these planes was over 30 years. This raised valid concerns about their reliability and effectiveness.

Furthermore, the delayed delivery of six new Boeing 737-MAX aircraft compounded Blue Air's problems. They had to continue using six older aircraft, which resulted in a 10% reduction in available seats and a 10% decrease in the use of the existing fleet. This caused maintenance costs to increase by EUR 25 million compared to 2019 (Reports of Cases Judgement of the General Court,2023).

In the first half of 2020, Blue Air saw a significant decline in operating revenue due to fewer flights and lower passenger numbers. This resulted in a net loss of EUR 28.29 million, which the Romanian Government compensated.

The decrease in flights and passengers was due to COVID-19 travel restrictions imposed by European states, which directly impacted over 400,000 passengers, with more than 250,000 having their flights cancelled (Reports of Cases Judgement of the General Court,2023).

The Romanian Government commissioned several reports to look at the impact of the withdrawal of Blue Air in the market. The reports showed Blue Air's withdrawal from the market, leaving many passengers stranded without affordable travel options, especially on routes where other low-cost airlines tend to use 'secondary' airports, which are further from city centres and considered less important than major airports.

European Commission orders Romanian Government to recover EUR 33.84 million in incompatible state aid from Blue Air

The European Commission conducted a thorough investigation into Blue Air's restructuring plan and concluded that immediate action must be taken to revise the plan to ensure the airline's long-term viability. The commission determined that the current plan is incompatible with the EU State aid rules. As a result, Romania must recover the EUR 33.84 million (RON 163.8 million) in illegal state aid awarded to Blue Air by the government (European Commission, 2024).

Following the report, the Romanian Ministry of Finance has been instructed to take swift action to recover the illegal state aid from Blue Air. The government spokesman, Mihai Constantin, has confirmed that the Ministry of Finance will take all necessary steps to recover the full amount from Blue Air, in accordance with the request from the European Commission.

Failure to do so may result in further legal action. The European Commission and the Romanian Government stated that they are committed to ensuring that all companies operating within the European Union respect the rules and regulations governing state aid.

Conclusion

Throughout its existence, Blue Air played a critical role in connecting small local entrepreneurs and the Romanian community living abroad with affordable air routes, which contributed significantly to Romania's regional economy.

Blue Air's commitment to providing affordable air travel to its passengers has been exemplary. The airline's exit from the market had significant consequences on the economy and more than 400,000 passengers, underscoring the importance of its services.

Although Blue Air's journey was not without its share of successes and failures, it has left a lasting legacy. The airline's commitment to providing affordable air travel to its passengers especially, small local entrepreneurs and the Romanian community abroad, has significantly impacted the economy positively. Blue Air's exit from the market has interrupted a vital service that would be difficult to replicate under the current circumstances.

Blue Air's journey has been remarkable, and its contribution to the Romanian aviation industry will always be remembered. Despite its challenges, the airline's legacy of providing affordable air travel to its passengers was a testament to its resilience and determination.

CHAPTER FOUR

The Failure of Flybe in the United Kingdom (UK): What Went Wrong?

Abstract

Flybe, once the largest independent regional airline in Europe, ceased operations on March 5, 2020, after failing due to various factors. These factors include high operating costs, fierce competition from low-cost carriers, and Brexit uncertainty.

Additionally, Flybe's business model, which relied heavily on short-haul domestic routes, made it vulnerable to external shocks, such as the COVID-19 pandemic.

The collapse of Flybe was expected to have significant implications for the UK's aviation industry. The company's failure has not only left a void in the regional airline market but also had immediate and long-term effects on individuals and communities. With Flybe's departure from the market, other regional airlines faced increased competition and pressure to fill the gap left by the company.

As a result, consumers saw higher prices for air travel in the short term. Additionally, the loss of Flybe led to a decrease in air connectivity for some regions, which could have negative implications for businesses and tourism in those areas.

This case study aims to provide a detailed analysis of the factors contributing to Flybe's downfall and delve deeper into these reasons.

Keywords: Business Model, Costs, Covid-19, Domestic Routes.

Providing Context to the Importance of Air Regional Connectivity in the UK

Air transport connectivity is an absolute necessity for regional population centres to position themselves in the world-city hierarchy and integrate into a global world (Burghouwt and Redondi, 2013). Despite the growth of low-cost carriers, UK regions are still limited in their ability to capture direct air services to intercontinental destinations due to their weak position in the UK urban hierarchy.

Although the existing literature on the retrospective economic and societal effects of air transportation in the UK is limited, previous research clearly demonstrates that airports and their air services can contribute to positive economic growth, boost Gross Domestic Product (GDP), create employment opportunities, and attract high-tech jobs (Laird and Mackie, 2018).

In 2018, the UK Department for transport analysed the economic impact of regional air services and concluded that the evidence supporting state aid was limited. However, governments worldwide continue to support routes where alternative modes of transportation are insufficient or inadequate to meet essential needs (Laird and Mackie, 2018).

It is essential to consider whether airlines like Flybe could have continued to operate while promoting alternative modes of transportation such as efficient rail connections, electric planes, green transportation, fast fibre availability, and the emergence of 5G technology. Further research is necessary to investigate the long-term sustainability of such airlines.

Background

Flybe was a prominent regional airline that operated in the UK, Ireland, the Channel Islands, and Europe. The company was founded in 1979 as Jersey European Airways and had its headquarters in Exeter, Devon. Flybe had its primary hub at Exeter International Airport, with additional hubs in Manchester, Belfast, and Birmingham. It operated a fleet of over 70 aircraft, serving more than 80 airports across the UK, Ireland and Europe (Flybe,1979-2020).

Flybe was known for providing vital connectivity in the UK regional market. Its strength was in the domestic UK regional market, which was defined as domestic flights excluding London routes and provided valuable connectivity to smaller airports, thereby contributing to the growth of local economies.

The airline was committed to serving remote destinations, particularly those separated from the rest of the country by water, where Flybe was sometimes the only realistic transport option.

The airline was known for making air travel accessible to a broader range of people, especially those who would otherwise have had to rely on lengthy and often unreliable road or sea transport. The airline had a unique passenger profile, with 50% of its passengers travelling for business, while the other 50% for leisure.

Tragically, Flybe collapsed on March 5, 2020, and was placed into administration. This left a trail of devastation, impacting many stakeholders and leaving no winners. The failure of Flybe resulted in 1000 job losses at airports, 800 job losses in the broader economy, and 2,300 Flybe employees losing their jobs (CAPA,2020).

These employees played a critical role in establishing Flybe as the UK's number four airline, the number one regional airline, and the biggest airline in the domestic UK region. The CEO of Virgin Connect (Flybe's new rebrand name), Mark Anderson, highlighted that Flybe has been achieving record operating results and customer service scores in recent months.

Therefore, it is highly probable that Flybe's failure was not due to the employee's inability to run the airline but was primarily caused by a history of commercial legacy problems, mainly financial, that had been plaguing the carrier for years.

Being a regional airline in the UK is a challenging business. A glance at a list of airlines that have been absorbed or failed in the UK in the past two decades suggests some potential successors of old that could have taken Flybe's domestic and possibly international routes, including Air Southwest, Brymon Airways, BMI regional/FlyBMI and Virgin Little Red. But they are no longer here! (Grant,2020).

Reasons for Flybe's failure

Despite repeated efforts to turn the company around, including a rescue plan in 2019, Flybe ultimately succumbed to the impact of the COVID-19 pandemic and ceased operations. This has raised legitimate concerns about the resilience of the regional airline market and its capacity to withstand future shocks.

Flybe's failure can be attributed to several factors, which this case study summarises in greater detail below.

Failing to make a profit

Flybe was listed on the stock market in 2010 and has struggled to achieve profit for several years. Despite carrying a significant number of passengers across its network of 24 airports in the UK, the airline had been unable to generate pre- or post-tax profits since 2010, which raised concerns about its long-term viability. This fact is supported by the company's annual report for 2017-18 and data from Companies House.

Moreover, Flybe has experienced persistent cumulative losses that have exceeded its profits over the nine years of its listing on the stock market. This has put the airline's future in question, leading to concerns among its stakeholders. In 2018 alone, Flybe carried 9 million passengers but still recorded an after-tax loss of £12.5m, highlighting its underlying financial challenges (Flybe Annual Report, 2017-18).

It's worth noting that Flybe's network included five airports that heavily rely on the airline for over 50% of their capacity. This information is shown in Figure 1, which ranks UK airports served by Flybe according to seat share. Overall, the airline's financial struggles have been a cause for concern among industry analysts and investors alike.

Figure 1: UK airports served, ranked by Flybe seat share at the airport, week of 15 June 2020

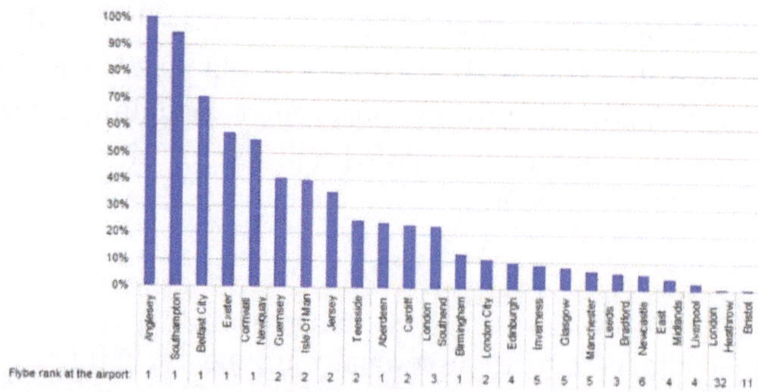

***Includes Jersey,Guernsey, Isle of Man**

Source: CAPA,2020

Flybe had been a key player for many airports in the UK, accounting for a substantial portion of their capacity, ranging from 10% to 50%. However, the airline's extensive network proved to be overly expansive, resulting in a record of losses that eventually led to its downfall.

One of the concerning issues with Flybe's operations was its monopoly on specific routes. Despite having a competitive advantage and a monopoly on some of these routes, the airline struggled to generate profits, as reported by CAPA in 2020. This situation raised concerns about the airline's impact on the overall air service offerings in the UK and the need for greater competition in the industry.

Another significant contributing factor to Flybe's financial struggles was its lack of access to London Heathrow Airport, which limited the airline's connectivity to international destinations through a hub. This limitation forced the airline to rely only on limited domestic routes, which affected its ability to attract a broader customer base. As highlighted by Rowland (2020), this lack of connectivity not only impacted Flybe but also had a significant impact on the overall air service offerings in the UK.

The collapse of Flybe resulted in a staggering 791,550 available seats being lost, which is a considerable loss for many airports in the UK. The airline's extensive network could have been more beneficial if managed more effectively, but its monopoly on specific routes had a detrimental effect on UK airports.

Nevertheless, despite facing financial challenges, the airline remained committed to delivering a quality travel experience to its passengers until it ceased operations in March 2020.

Arrears in the payment of Airline Passenger Duty (APD) charges

The aviation industry has been facing challenges since 1993, when the Air Passenger Duty (APD) was introduced. Many aviation industry stakeholders believed that the APD would hinder their efforts to decrease air travel costs. However, the opposition against APD in the domestic market was challenging, mainly because aviation fuel remained untaxed. This may partly explain why the cost of a Flybe flight from Southampton to

Manchester on a Monday morning was only £99, while the train fare was £236 (as of the time of publication).

According to Hockley (2020), short-haul aviation is significantly undertaxed due to the externalities associated with short-haul air travel, the demographic profile of air travel passengers, and the lack of empirical evidence supporting the developmental value of short-haul routes compared to investments in rail, road, and broadband infrastructure. However, the situation has changed now, and APD is now considered a significant obstacle for all domestic and international carriers.

The UK APD is the highest tax of its kind in Europe and among the highest in the world (Seetram et al., 2013). Flybe's woes were exacerbated by a double APD charge, which inflated several UK domestic route's airfares. The APD was levied per sector, which differed from international flights, where it is charged only on the outbound journey (Seetram et al., 2013).

When Virgin Connect acquired Flybe, it had arrears of up to £106m in APD tax. It is unclear why Flybe had APD tax arrears, considering that APD is collected at the point of sale, i.e., once the passenger has paid and the air ticket has been issued. The rapid accumulation of APD arrears suggests that Flybe had a flawed business plan with no accountability (Flottau, 2020).

Despite the government's promises to review the APD, the then-Chancellor's budget presentation for 2020 did not mention any changes to the APD at that time (Flottau, 2020). The aviation industry continues to face challenges related to APD, and stakeholders are looking for ways to mitigate its impact on their operations.

The Impact of Covid-19 on Flybe

The Covid-19 pandemic has significantly impacted the airline industry globally, resulting in reduced passenger demand and substantial financial losses for airlines.

According to a previous study conducted by the World Travel and Tourism Council, the pandemic was predicted to cause a decrease in global travel of up to 25% in 2020. To put this into perspective, the decrease is equivalent to losing three months of international travel (Gossling et al. 2021).

Flybe was one of the major casualties of the outbreak. The impact of the Coronavirus on flight bookings was seen as the final straw that broke the camel's back. In his farewell message to his staff, Flybe's Chief Executive, Mark Anderson, stated that the Coronavirus had "put additional pressure on an already difficult situation."

Aircraft Fleet Challenges

On January 13th, 2020, Flybe proudly held the distinction of being the fourth largest airline in the UK, having an impressive fleet of 67 aircraft in service. This information was obtained from the CAPA Fleet Database (Figure 2), where Flybe was placed behind British Airways, EasyJet, and Jet2.com but ahead of comparable operators such as TUI Airways.

These statistics reflect Flybe's exceptional credibility and noteworthy success in the aviation industry, which has made it one of the most reliable and trustworthy airlines.

Figure 2 : UK airlines* ranked by total number of aircraft in service, 13-Jan 2020

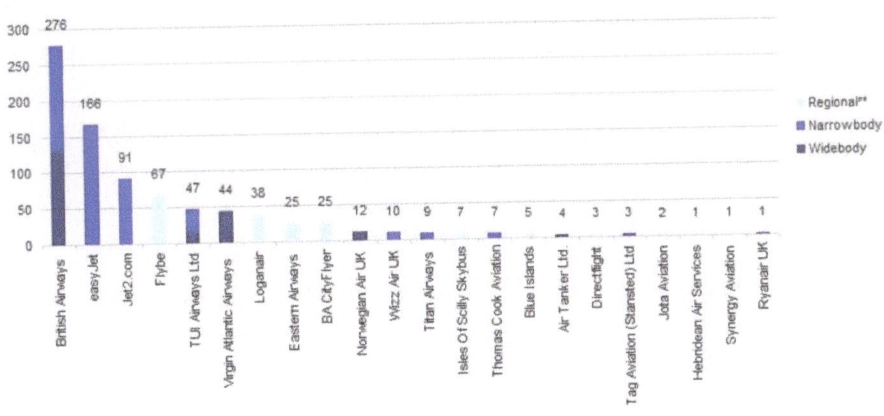

**Only airlines that operate passenger aircraft are included.*

*** Regional aircraft, including small turboprops.*

Source: CAPA Fleet Database (2020)

However industry experts have identified one of the key reasons behind Flybe's financial struggles - their aircraft fleet was too large for the routes they operated on, leading to inflated operating costs.

This issue likely arose due to Flybe's expansion strategy in previous years, which involved acquiring larger aircraft to accommodate growing passenger numbers. However, as the airline's financial situation worsened, the oversized fleet significantly burdened their finances.

This problem was further compounded by Flybe's operation on regional routes with lower passenger volumes, making it difficult for them to fill up the larger planes and achieve economies of scale.

Lack of Service Frequency on Some Routes

Flybe faced significant challenges due to its infrequent services on many routes, which made it difficult to compete with larger airlines that offered more frequent flights. In spite of these challenges, Flybe's slow capacity growth rate during the winter of 2019/2020 indicates that the airline started taking a cautious approach to its development plans in the face of a challenging global economic environment.

According to a report by CAPA (Centre for Aviation), Flybe's cautious approach reflected the airline's efforts to manage its resources and costs better, while also focusing on improving its operational efficiency and customer experience.

Conflicting Shareholder Objectives

In early 2019, Flybe was acquired by a consortium known as Connect, which included Virgin Atlantic, the Stobart Group, and Cyrus Capital. As part of the acquisition, the consortium pledged to invest £100 million into the ailing airline, which was struggling with mounting debts and fierce competition in the market. However, soon after the acquisition, differences in shareholder objectives emerged, with Flybe's shareholders having divergent goals (Flottau,2020).

While Virgin Atlantic was primarily interested in providing a feeder service for passengers to connect with their long-haul flights at Heathrow and Manchester airports, the acquisition offered an opportunity to run more regional services and increase passenger traffic at Southend Airport, which Stobart owned (Lawrie,2020).

Despite the pressing need to address Flybe's fundamental challenges, the consortium provided a £2.2 million rescue package to keep the airline afloat. However, investors only received 1 pence per share, despite the shares being floated at 295p in 2010, resulting in significant losses for investors. This move was deemed necessary to safeguard the airline's long-term future after mounting debts impacted it, as stated in the Flybe's Annual Report 2017-18.

Critics on Connect's acquisition of Flybe believe that Richard Branson, the founder of Virgin Atlantic, had been allowed to revive his ambitions to operate a UK domestic airline following the failure of Virgin's Little Red subsidiary in 2014. However, Connect maintained that the move was prudent and in the best interest of the airline's long-term future.

Conclusion

The UK's urban hierarchy has been undergoing significant changes in recent decades, necessitating strong regional air connectivity for economic growth, which was a service provided by Flybe. The collapse of Flybe, the fourth UK airline to go out of business in two years, has cast a spotlight on the urgent need for the aviation industry's long-overdue reforms.

Despite transport ministers promises since 2018, the necessary reforms have yet to be implemented, leaving the industry and its stakeholders in a precarious position. According to Rose (2014), the challenges in the aviation sector in the UK have been evident for some time, and the Flybe collapse is further evidence of the need for wholesale reform.

The UK government's decision to not intervene in Flybe's collapse has raised serious concerns about their commitment to improving regional connectivity. However, in January 2020, the UK government provided a rescue package to save Flybe from collapsing, which has led some to speculate that the UK government's efforts to save Flybe were politically motivated and aimed at gaining support in Northern England, where the Conservatives had pledged to improve regional connectivity in their manifesto (Topham, 2020).

The collapse of Flybe highlights the importance of strategic decision-making and adaptability in the aviation industry. Airlines must carefully evaluate their routes, fleet size, pricing, and frequency of services to remain competitive and profitable in an ever-changing market. The UK aviation industry's reforms must also be implemented, as promised by the government, to improve regional connectivity and ensure the sector's long-term sustainability.

CHAPTER FIVE

Flybe Grounded by Second Administration in 2023

Abstract

Flybe 2.0 resumed operations in March 2022 after going bankrupt in March 2020. The airline aimed to become the UK's leading regional airline by prioritising safety, operational excellence, customer service, teamwork, and learning from past mistakes.

However, on January 28, 2023, the airline entered administration for the second time, causing significant disruption to passengers and stakeholders.

Around 300 flights scheduled to operate on 17 different routes were affected, impacting 22,776 departure seats across the airline's network. 76% of the carrier's capacity was on domestic UK routes.

Ultimately, Flybe 2.0 collapse resulted from a combination of factors, including operational performance issues, financial difficulties, and an overly ambitious expansion plan.

It is now more critical than ever that the aviation industry examine these issues closely and work towards finding viable solutions. The collapse of Flybe 2.0 should serve as a wake-up call for the industry, and it must work together to ensure that it creates a more sustainable, flexible, and resilient aviation industry for all.

This case study will examine the root causes that led to the collapse of Flybe 2.0 and reflect on whether the airline's brief revival was a temporary fix or if its leadership failed to adapt to its challenges.

Keywords: Administration, Capacity, Collapse, Stakeholders.

Background

Flybe 2.0, a formerly defunct airline, made a remarkable comeback in April 2021, having secured an Air Operator Licence (AOL) and type A and B route licences from the UK Civil Aviation Authority (CAA). With these licenses in place, the airline was authorised to operate both schedules and charter passenger services, paving the way for a successful comeback in the aviation industry (CAA, 2023).

Towards the end of April 2021, Flybe 2.0 leased 86 remedy slots from British Airways in a strategic and well-planned move. These slots enabled the airline to operate services from Heathrow to Aberdeen (18 weekly slots) and Edinburgh (25 weekly slots), primarily for the summer season until the end of October 2021 (Calder, 2022).

This move significantly boosted Flybe 2.0's operations and revenue, as Heathrow is one of the busiest airports in the world and a key hub for international travel. Additionally, Flybe also received permission to operate in Birmingham and Manchester, further strengthening the airline's position in the UK aviation market. Despite extensive planning for the re-launch, the venture launch had to be delayed due to the ongoing coronavirus-related restrictions during the summer of 2021.

The number of UK domestic flights in 2021 had decreased by almost 60% compared to pre-pandemic levels in 2019, significantly impacting all airlines, including Flybe 2.0. However, Flybe 2.0 boldly initiated ticket sales on March 20, 2022, and announced Belfast City Airport as its second base. The airline confidently offered 23 routes with up to 230 rotations per week for summer 2022, covering 16 airports across the UK, France, and the Netherlands, all under the new motto, "Smile and go the extra mile" (CAA, 2023).

Interestingly, the airline faced no direct competition from any other airline on at least eight routes. The airline's inaugural flight from Birmingham to Belfast took off on April 13, 2022, with much fanfare, followed by six more routes in the same month, including Belfast-Glasgow, Birmingham-Amsterdam, and Heathrow-Amsterdam. Moreover, Flybe 2.0 announced plans to add 13 more sectors during July and August 2022, serving cities such as Inverness, Southampton, Newcastle, and Avignon, thereby establishing its dominance in the market (Joint Administrators progress report,2023).

The airline's return to the skies was met with great enthusiasm from aviation enthusiasts and the industry alike, making it clear that Flybe 2.0 means business and is here to stay.

Unfortunately, on January 28, 2023, at 3 a.m., Flybe 2.0 went into administration. As a result, all aircraft were grounded in the UK, and many pilots and crew members were laid off. In total, 234 redundancies were made immediately, and 45 employees were retained to provide a stable platform and maintain the required postholder roles to allow sufficient time to pursue a potential business sale out of administration.

John Pike and Michael Robert Pink from Interpath Advisory were appointed by the High Court as Joint Administrators to oversee the affairs, business, and property of Flybe 2.0 Limited on the 28th of January 2023.

The primary objective of the administrators was to achieve a superior outcome for the company's creditors collectively, compared to the alternative of winding up the company. It is worth noting that the granted order is valid until January 25, 2025, giving the company some time to explore potential business sale opportunities (Joint Administrators Progress report, 2023).

Why did Flybe 2.0 fail for the second time?

Flybe 2.0 collapsed due to a combination of several factors that had a detrimental impact on its overall business health. One key reason behind its downfall was an overambitious expansion plan, which involved overstretching its operations and resources, leading to a significant increase in costs.

Additionally, the airline faced operational performance issues, including flight delays, and cancellations, which resulted in a decline in its reputation and customer loyalty. Moreover, Flybe 2.0 was grappling with severe financial difficulties, including mounting debts, high overheads, and fierce competition from other low-cost airlines. All these factors combined created a perfect storm that ultimately led to the airline's downfall.

Repeating past mistakes – Aircraft lease challenges

One of the major problems the airline faced was that it did not own the aircraft it operated. Instead, all of its aircraft were leased from NAC or

Aergo Capital. When Flybe 2.0 collapsed, all of the leases had already been terminated by the lessors. This meant that the aircraft were not Flybe 2.0's assets to release, and the airline could not recover any value from them (CAA, 2023)

Flybe 2.0 had signed leases for five additional aircraft in April 2022, with the expectation of receiving 13 out of 17 by June 1, 2022. However, the airline delayed in restarting its operations until March 2022 and had only two aircraft in service at that time. This delay, combined with frequent cancellations and long delays, severely impacted the airline's operational performance and reputation. Flybe 2.0's inability to acquire more aircraft further worsened the situation and severely impacted its capacity to expand and operate the intended number of routes (Perry, 2023).

One of Flybe 2.0's goals was to expand and offer feeder traffic on international routes through regional services. However, this plan had to be revised due to the airline's operational performance issues and financial difficulties. The airline aimed to increase load factors for domestic flights while attracting long-haul passengers to the US and Europe. However, the plan was overly ambitious and not realistic given the airline's current situation.

During the summer schedule of 2022, Flybe 2.0 had to cancel up to 700 flights until the end of October. The Civil Aviation Authority (CAA) report indicated that the cancellation rates for flights departing from Edinburgh and Glasgow were approximately 12% and 20%, respectively. These cancellations caused considerable inconvenience to Flybe 2.0's customers and further damaged the airline's reputation (CAA, 2023)

Flybe 2.0's fleet expansion plan aimed to increase its aircraft fleet to 32 within a short period of time. On the other hand the expansion led to financial difficulties for the company. The airline's inability to acquire more aircraft severely impacted its capacity to expand and operate the intended number of routes, further worsening the situation.

A decrease in demand and an increase in delays resulted in a loss of approximately £30 million in revenue and additional expenses for the airline. It is difficult to understand how increasing the fleet threefold would have improved the financial situation, especially given the low demand caused by the pandemic (CAA, 2023).

Contractual Issues

The aviation industry is no stranger to the challenges of managing aircraft contracts, which can lead to issues with performance and delivery. One of the biggest obstacle stems from the pre-signature stage, where unrealistic schedules are often established, leading to over-commitment. This put companies in a difficult position where they must still meet their commitments, often resulting in penalties if they don't (Perry,2023).

Flybe 2.0 faced significant difficulties in managing its aircraft engines under long-term lease agreements, according to industry reports. This issue became increasingly challenging when the company collapsed, and it needed to release capital to pay its creditors.

During the administration process, the administrators sold ten of the airline's unfinanced engines for £2.4 million. However, the remaining leased engines, valued at £85 million, had to be returned to their respective companies (Joint Administrators progress report,2023).

To release the value of the company's inventory, IBA Group Limited, an independent aviation agent/valuer, was appointed to conduct a thorough desktop valuation. This valuation was conducted to develop a comprehensive realisation strategy for all aircraft parts and tools.

While the administrators successfully sold some of the company's stockholding, they also pursued potential counterparty litigation matters on the contracts (Joint Administrators progress report,2023).

Financial Woes of Flybe 2.0

Flybe 2.0 was a struggling airline that faced significant financial difficulties. Despite selling off its assets and paying its preferred creditors, the company still had an estimated debt of £82,612,688. To make matters worse, Flybe 2.0 was losing between £4 million to £5 million on average each month, putting immense pressure on the company's finances (Flybe Limited, 2023).

To keep the airline afloat, Flybe 2.0 had to draw £29 million from the increased Revolving Credit Facility (RCF). However, due to rising costs, the RCF had to be increased from £5 million to £33 million, which was an alarming development for any commercial airline operator (Flybe Limited, 2023).

When it came to assets, Flybe 2.0 had very little of value left apart from its remaining cash reserves. The only assets of substantial intrinsic value were the Civil Aviation Licenses and the slots, which were subject to contractual restrictions that affected their transfer. This meant that selling these assets to raise additional funds was not a viable option for Flybe 2.0.

Last roll of the dice to save the Airline

As Flybe 2.0 faced the prospect of going out of business, the company's administrators made a desperate attempt to rescue it. They applied to the UK CAA for a temporary operator's license, hoping to restructure the business and prolong its lifespan.

To achieve this goal, the administrator engaged with key stakeholders to acquire the business, obtain the Aircraft Operating License (AOL) from the CAA, and protect the landing slots at Heathrow and Amsterdam through Airport Coordination Limited (ACL) and Netherlands (ACNL). However, the company's administrators encountered several obstacles in its efforts to retain its slots and extract value from them, particularly in the absence of a sale or substantial transfer of the company's operations.

The most significant challenge was that the prized Heathrow slots, consisting of seven pairs of take-off and landing slots, did not belong to Flybe and could not be transferred or sold by the company. Flybe had obtained these slots through an agreement with British Airways (BA), which had been mandated by the European Commission. Under the agreement, if the slots were at risk of being forfeited due to the "use it or lose it" rule, they had to be surrendered to BA unless a buyer for the company emerged (Joint Administrators progress report, 2023).

Another area for improvement was the recognition of a temporary license allowing Flybe to operate in French and Swiss locations by European authorities, given that the UK was no longer a member of the European Union. Despite the administrator's efforts, no offers were made for the slots, and they were returned to the pool by the relevant slot coordinators. Although Flybe's role in the market was smaller than it had been during its previous collapse in March 2020, the rescue effort proved to be a complex undertaking.

Conclusion

The failure of Flybe 2.0 has had a significant impact not only on the company's employees and shareholders but has also hit the aviation industry hard, leading to increased uncertainty and instability. This news came as a surprise to many, especially considering that the company had received significant financial backing from wealthy investors with billions in assets.

Experts in the aviation industry had already cautioned against the unrealistic targets set by Flybe's previous leadership. Unfortunately, the company's goals failed to account for the impact of the COVID-19 pandemic on its business and the subsequent decline in bookings. As a result, Flybe's financial situation deteriorated.

The collapse of Flybe 2.0 indicates the underlying issues that have plagued the industry for years. It is essential that the industry examine these issues closely and work towards finding viable solutions to create a more sustainable and resilient aviation industry for the future.

CHAPTER SIX

The Fall of a Giant - Indian Airline Jet Airways

Abstract

The aviation industry has witnessed significant transformations in recent years, driven by diverse commercial models and technological advancements. These changes have brought about both opportunities and challenges in equal measure, particularly for the Indian aviation sector, which is yet to confront the challenges that lie ahead effectively.

A case in point is Jet Airways, one of India's largest private airlines, that faced a severe shortage of funds, which forced it to temporarily halt its operations .Despite gaining a substantial market share and revenues, Jet Airways gradually became a loss-making enterprise, leading to its eventual downfall on April 17, 2019, as it needed more funding to sustain its operations.

This case study highlights the challenges faced by Jet Airways including fierce competition, rising fuel prices,and regulatory hurdles. These factors have made it increasingly more work for Jet Airways to sustain profitability, particularly in the face of mounting debt and operational costs.

Keywords: Competition, Commercial Models, Profitability, Stakeholders.

Background

It's important to understand the context behind the suspension of operations and eventual bankruptcy of Jet Airways, one of India's oldest private airlines. To do so, let's look closely at the company's history.

Jet Airways was founded in the early 1990s by Naresh Goyal, who began his career in the aviation industry as a ticketing agent.

In 1992, the Indian government opened the domestic air market to allow private air carriers to operate domestic flights under the Air Transport Operator (ATO) license (Shastri,2014).

Jet Airways was one of the carriers that emerged during this license rush, with a high-profile launch. However, as a startup airline, Jet Airways was required to comply with government traffic allocation regulations, which included adding capacity on unprofitable regional routes in Northeast India.

On April 18, 1993, Jet Airways received its first two aircraft. The request for their acquisition came from Naresh Goyal and JRD Tata, widely recognised as the "Father of Indian Aviation. In the early years of Jet Airways, JRD Tata entrusted Naresh Goyal with the responsibility of leading the company, challenging him to make Jet Airways better than the best. He famously stated, "Naresh, if you cannot make Jet Airways better than the best, then send these two aircraft back today."

In 2003, India's aviation industry witnessed a significant transformation by introducing its first Low-Cost Carrier (LCC), Air Deccan. This event marked the entry of several private LCCs, including GoAir, King Fisher, and SpiceJet, which increased the industry's overall capacity and led to lower airfares.

In response to this change, carriers adopted different strategies to gain market share. For example, LCCs made air travel more affordable by offering lower fares to the middle class, thereby increasing demand (Doganis, 2001; O'Connell and Williams, 2006). As a result, Full-Scheduled Carriers (FSCs) such as Jet Airways discounted their airfares by up to 70% on some routes to compete with the LCCs.

This reaction is common in the airline industry. While an aggressive pricing strategy may benefit consumers and increase demand in India's metro and tier 2 cities, the opposite effect was witnessed in the form of low revenues, low margins, and loss of market share, ultimately impacting profitability for airlines (Krishnan, 2008).

By the end of 2003, Jet Airways had a fleet of 41 aircraft and was operating over 250 flights daily. It was listed on the Bombay Stock Exchange and became a public company on December 28, 2004. Between 2004 and 2005, the Indian aviation industry underwent a significant shift.

Indian carriers, including Jet Airways, received a notable increase in aircraft orders. This shift was a clear indication of the liberalising effects of political change, as the Indian Government began to loosen their grip on monopolization and remove restrictions on regulatory barriers in key industries.

Jet Airways demonstrated its adaptability and commitment to economic development by capitalising on this opportunity (Kathpal and Akhtar, 2021). Jet Airways' fleet, with an average age of 4.45 years, reflected the company's commitment to modernisation.

The airline had established an impeccable reputation for punctuality and excellent service, making it a trusted choice among business passengers. This reputation elevated Jet Airways to a distinguished status in the Indian aviation industry, instilling confidence in its reliability and service quality. As a result, Jet Airways was hailed as India's most respected private airline.

Jet Airways operated flights to 65 destinations, including major cities such as Abu Dhabi, Bahrain, Bangkok, Brussels, Colombo, Dubai, Hong Kong, Jeddah, Kuala Lumpur, Kuwait, London, Muscat, New York, Singapore, and Toronto. Its extensive fleet of over 85 aircraft comprised 12 Airbus 330-200, 20 ATR 72-500, 11 Boeing 737-700, and 42 Boeing 737-800, covering approximately 1,000 routes (Ghosh, 2019).

The airline had ambitious plans, including developing its own maintenance hangars and pilot training centres to provide the best possible experience to its passengers. Jet Airways was renowned for its unyielding commitment to safety, evidenced by its well-maintained fleet and highly trained crew.

Jet Airways' market share dropped to one-third in 2007 from controlling about 50% of the domestic market in 2003 due to overcapacity, aggressive expansion of low-cost carriers (LCCs), and an inability to control costs. The company then experienced a slow and painful descent from being India's largest airline in 2010 (based on market share) to the company posting net losses for seven years out of the last nine financial years (2010-2019) (Kathpal and Akhtar, 2021).

Jet Airways ceased its operations on April 17, 2019, due to a severe cash shortage and the inability to obtain emergency bank loans. As a result, thousands of employees were left without payment for their work in March and April of 2019.

Furthermore, the airline defaulted on loans and payments to vendors and lessors, eventually leading to its bankruptcy in the Netherlands on May 21, 2019. Following this, two Indian creditors filed bankruptcy petitions in the Indian court on behalf of twenty-six lenders on June 17, 2019 (ibef.org; PLR,2019).

In retrospect, the consortium lenders and other industry experts have raised several questions about Jet Airways' journey. These include inquiries into what went amiss over the years, whether it was a result of poor financial management, corporate governance issues, or the intense competition within the Indian airline industry, which Jet Airways could not handle.

The above-mentioned effects inevitably lead carriers to one of two outcomes: cessation of operations, acquisition, or merger with other airlines or business investors. In the case of Jet Airways, the answer would have been to combine all these possibilities and more.

Reasons for Jet Airways Down Fall

The Writing was on the Wall – Unsustainable Financial Losses.

Jet Airways experienced financial difficulties in March 2018, when it incurred losses of a staggering amount of INR 1036 crore due to various factors such as rising fuel prices, low fares, and increased competition from other airlines. These losses were just the beginning of the airline's woes, as it faced severe challenges in the months that followed (Arora & Ravi, 2019).

In August 2018, Jet Airways faced a severe financial crisis after deferring the reporting of its second-quarter year earnings and reducing employee salaries by 25%. This move prompted the Indian Directorate General of Civil Aviation to conduct a financial audit of the company, further highlighting its precarious financial position. During the same month, Jet Airways posted a staggering loss of Rs. 1323 crore, approximately equivalent to USD 18 million (Sun, 2021).

The airline's troubles continued in September 2018 when the Income Tax Department launched a survey in the Delhi and Mumbai offices of Jet Airways, which accused the airline of financial misappropriation of funds. The surveys aimed to investigate the company's financial records and transactions and assess if any discrepancies or financial misappropriation had occurred. These accusations dealt another blow to the company's already damaged reputation (Ghosh, 2019).

In the fiscal year 2019, Jet Airways reported a loss before tax of 55.36 billion Indian rupees (approximately USD 661,829,296 million). After a series of unprofitable years, a revenue decline and a significant cost increase occurred during the financial year 2014-2019 (shown in Figure 1). Subsequently, Jet Airways was forced to curtail some of its operations, including the non-payment of salaries to some of its employees, citing "circumstances beyond its control" (Ghosh, 2019). These actions further exacerbated the airline's financial crisis and significantly reduced its overall operations.

These actions further exacerbated the airline's financial crisis and led to a significant reduction in its overall operations. In summary, Jet Airways's financial troubles began in 2018 and continued for several years, leading to a decline in its reputation and operations.

The airline continued to face challenges in September 2018, when the Income Tax Department launched a survey in Jet Airways' Delhi and Mumbai offices, alleging financial misappropriation by the company. The surveys were intended to investigate the company's financial records and transactions and assess whether any discrepancies or financial misappropriation had occurred. These accusations dealt another blow to the company's already damaged reputation (Ghosh, 2019).

Figure 1. Key financial performance of Jet Airways Limited from financial year 2014-2019 (in Indian rupees)

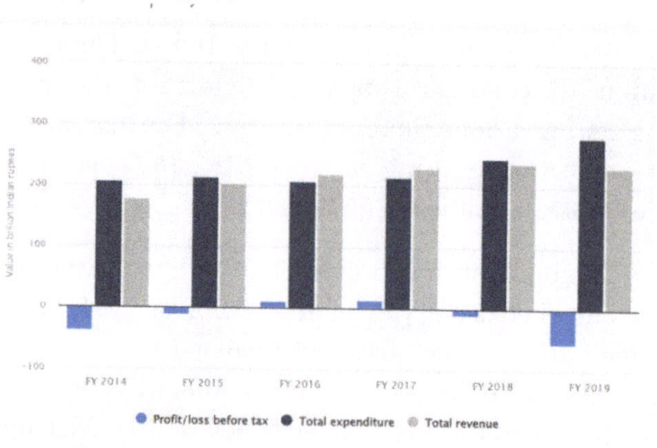

Source: Statista, 2024

Bad Management Decisions: The Acquisition of Air Sahara

Mergers and Acquisitions (M&A) are considered to be essential strategic actions that enable a company to achieve external growth, gain a competitive advantage, and improve its overall financial performance.

In the current global economy, M&A is increasingly utilised worldwide to enhance a company's competitiveness by increasing its market share, diversifying its portfolio to minimize business risk, expanding into new markets and geographies, and capitalizing on economies of scale. These actions help companies achieve their long-term business goals, enhance shareholder value, and stay ahead of the competition (Daddikar and Shaikh 2014).

In 2007, Jet Airways made a strategic decision to enter the Low-Cost Carrier (LCC) market by acquiring Air Sahara on April 12, 2007, for a total valuation of approximately USD 332.61 million (Ghosh,2018). However, despite the benefits that M&A can bring, Jet Airways's acquisition of Air Sahara turned out to be a questionable move that had a significant impact on the company's financial performance (Mahalakshmi, 2013).

Many observers argue that Jet's financial problems started when Goyal bought out its rival, Air Sahara. This deal gave Jet Airways endless problems – economic, legal, and human resource-related.

Goyal bought Sahara to take on Kingfisher Airlines and low-fare carriers Air Deccan, IndiGo, and SpiceJet. However, the valuation of Air Sahara was considered a "significant overvaluation" by many experts from the industry, and the deal reduced Jet Airways's ability to spend extra money to take on the competition effectively. Jet Airways wasted the Initial Public Offering (IPO) funds that remained after placing plane orders (Mahalakshmi, 2013).

It is important to mention that, at the time of the acquisition, Air Sahara had already accumulated losses of around USD 160.57 million (Mahalakshmi, 2013). The acquisition was financed from additional borrowings, which increased Jet Airways's already high leverage. Moreover, Jet's management could have lapsed in due diligence, which would have cost them dearly in the years to come (see Table 1 some of the impacts of this acquisition).

In summary, while M&A is an effective way to achieve business growth and compete in the market, companies should be cautious in their approach and conduct thorough due diligence to ensure that their investment decisions are sound and aligned with their long-term goals.

Table 1. Acquisition of Air Sahara by Jet Airways.

Benefits Identified by Jet Airways	Impact
Air Sahara's Fleet of 27 Aircraft (Saran & Mehra, 2012)	Air Sahara had a fleet of 27 aircraft, whereas Jet Airways had a mixed fleet of different-sized aeroplanes, resulting in higher maintenance costs.
New international routes, infrastructure and logistics (Saran & Sahgal, 2006)	Jet Airways had to renegotiate the terms with the operators to get access to airport facilities for new international routes, infrastructure, and logistics.
Manpower	The two airlines had very different operational environments, with Air Sahara offering higher remuneration to its personnel than Jet Airways. This became an issue when the two workforces were merged.
Economies of scale	Jet Airways operated Jet Lite as a separate entity and needed to merge its operations fully, failing to realize the full potential of economies of scale.
Brand recognition	The lack of brand separation between Jet Airways and Jet Lite on their domestic routes was a significant oversight that should have been addressed to avoid confusion among customers.

Source: Author and referenced sources

Aggressive Growth Strategy

Jet Airways, implemented an ambitious growth strategy that resulted in significant expansion. However, this expansion coincided with the global economic downturn, which adversely affected the airline's profitability. In April 2018, Jet Airways proposed a merger with its subsidiary, Jet Lite, with the goal of streamlining operations and reducing expenses. Unfortunately, this proposal was rejected by the Indian government, resulting in a further decline in the airline's financial situation.

Jet Airways subsequently introduced a Revenue Enhancement Programme to achieve 3-4% growth in revenue per available seat per kilometre (RASK). The programme included tactical and strategic initiatives involving network, pricing, inventory management, and sales. Despite these measures, Jet Airways continued to incur losses, with reports indicating a further decline of INR 1323 crore (equivalent to USD 180 million) during the April-June 2018 period (Ghosh, 2019).

Given the airline's prominence and significance in the aviation industry, these developments were particularly concerning. The events leading to Jet Airways's downfall serve as a warning to companies about the criticality of responsible financial management in the aviation industry. Companies must take appropriate measures to manage their finances effectively and ensure their long-term sustainability.

Bad Management Decisions on Choice of Aircraft Fleet

The aviation industry is recognised for its intense competition, where airlines must consistently make informed decisions to maintain a leading position in the market.

In 2012, Kingfisher, one of India's prominent airlines, ceased its operations, resulting in a significant gap in the market. Subsequently, IndiGo, a low-cost airline, quickly emerged as an industry leader, taking over Kingfisher's position and toppling Jet Airways, which was already struggling to sustain its operations.

Jet Airways decided to counter this competitor threat by purchasing a mixed fleet of ten wide-bodied Airbus and Boeing 777 aeroplanes. However, this decision could not have been practical and thought through, as it increased the cost of the resources required to maintain and operate the mixed fleet.

Jet Airways configured these aircraft with only 308 seats, which was much lower than the global standard of about 400 seats, resulting in a notable loss of potential revenue (Economic Times, 2023).

Even after changing the configuration to 348 seats, Jet Airways retained eight unprofitable first-class seats, despite being advised by the then commerce chief and longtime associate of the Chief Executive Officer. These seats added an unnecessary eight tonnes to the aircraft's weight and earned nothing for the flight. Furthermore, no proper network was ever planned for the aircraft, which resulted in further losses for Jet Airways (Economic Times, 2023).

Conclusion

The Indian aviation industry has undergone significant transformations over recent years, and Jet Airways presents a case study of how an airline can go bankrupt despite earning substantial revenue and having a high market share. The airline's lack of planning and foresight ultimately contributed to its downfall.

Jet Airways experienced losses for several years, and struggled to adapt to the changing market dynamics brought about by the emergence of LCCs. Consequently, it faced high operational costs and minimal profits until it ceased operations in 2019.

Numerous industry experts attribute Jet Airways founder, Mr Naresh Goyal, to questionable strategic decisions and an obsession with poor choices in mergers and acquisitions. While investors are currently planning to revive the airline, it is essential to note that the Indian aviation industry has undergone significant changes since the airline's departure.

The emergence of LCCs has disrupted the market, and the government has implemented economic liberalisation and regulatory reforms that have increased the number of aircraft operating within India and boosted inbound and outbound tourism. Nevertheless, while these reforms have been helpful, more measures are necessary to build a resilient and sustainable aviation industry in India.

To achieve this goal, the Indian government must develop a longer-term, multi-faceted strategy to support and protect airlines from failing without impacting taxpayers. This strategy could involve measures such as creating a fund to help struggling airlines, developing a regulatory framework that encourages competition while ensuring the safety and security of passengers, and providing incentives for airlines to invest in new technologies and sustainable practices.

Ultimately, a more comprehensive approach to reform is essential to build a resilient and sustainable aviation industry in India. This will require collaboration between the government, airlines, and other stakeholders to address the industry's challenges and create a brighter future for aviation in India.

In conclusion, the aviation industry is highly competitive, and airlines must make informed decisions to stay ahead. Jet Airways failed to take calculated risks, which resulted in its downfall. Therefore, airlines must consider global standards and make decisions in the best interests of the company, its employees, and its customers.

CHAPTER SEVEN

The Failure of an Iconic Brand, Thomas Cook - Never Too Big to Fail.

Abstract

This case study provides a detailed analysis of the unfortunate downfall of Thomas Cook, one of the largest travel brands worldwide. The case study examines the factors that led to the company's failure, including internal and external influences.

The study delves deeply into the critical reasons for Thomas Cook's collapse, exploring aspects such as its management practices, business model, and financial performance. It further evaluates how external factors such as economic downturns, geopolitical uncertainties, and changes in consumer behaviour played a significant role in the company's demise.

The case study further provides insights into the company's history, its market position, and its growth trajectory, thereby offering a holistic understanding of Thomas Cook's journey. It provides an in-depth look at the company's strategic decisions.

In conclusion, the study identifies a combination of factors that led to Thomas Cook's unfortunate demise. While external factors played a significant role in the company's collapse, the study highlights how internal factors, such as poor management decisions, contributed to its downfall. The case study additionally offers valuable lessons for businesses in the travel industry and beyond, highlighting the importance of being agile and responsive to changing market conditions.

Keywords: Collapse, Debt Issues, External Influences, Salaries.

Background

Tour operators play a significant role in the travel and tourism industry by promoting and developing tourist destinations. Several studies have emphasised the highly competitive nature of the tour operating business in the United Kingdom (UK), characterised by price wars, frequent appearances and disappearances, takeovers, and mergers of different tour operator businesses (Schwartz et al.2009).

This industry is particularly competitive in the UK, where the top five operators, including Thomas Cook, control around 60% of the package tour business, a trend that has persisted since 1986. Despite this oligopolistic structure, industry experts believe the tour operating business remains competitive (Leadbeater, 2019).

Over the years, British and international tour operators have adopted various strategies to enhance their profitability, including vertical integration strategies combining their tour operator, travel agent, and airline business. This strategy aims to improve profitability and increase market power by controlling the distribution channel.

As the larger tour operating businesses aggressively expanded, smaller tour operators faced several challenges, such as access to distribution channels, negotiating competitive terms with accommodation providers, access to air capacity, and the general competitive threat posed by larger competitors (Economic Investigations, 2015).

Overall, the tour operating business is a complex and competitive industry that requires businesses to be strategic and innovative to succeed. While the market is highly concentrated, there is still room for smaller operators to thrive by adopting unique strategies that differentiate them from their larger counterparts.

Who was Thomas Cook?

Thomas Cook is a name that has become synonymous with modern package holidays, despite its long and storied heritage. Born in the market town of Melbourne in Derbyshire in 1808, Thomas Cook began his career as a cabinet maker before later becoming a Baptist Minister. He had a mission to make travel simple, easy and pleasurable for everyone. He is widely regarded as a 19th-century tourism entrepreneur and teetotal social reformer, as evidenced by many research papers (Leadbeater, 2019).

In 1841, Thomas Cook organized his first day tour from Leicester to Loughborough. Seven years later, he arranged a day trip for 90 passengers from Humberstone Gate, Leicester, to Belvoir Castle, which was set off in a seven-horse-drawn carriage.

Thomas Cook's global reputation and financial strength allowed him to establish relationships with hundreds of small tour operators, hotels, and caterers, making him a master of package holidays (Sims, 2019).

Overview of why Thomas Cook failed

On September 23rd, 2019, at 2:00 am, Thomas Cook, the iconic British travel company, announced its collapse after 178 years of operation.

Approximately 150,000 British citizens were stranded abroad, and hundreds of thousands of people from other countries who had also been travelling with Thomas Cook found themselves in a similar predicament (McCormick,2019).

The situation was further complicated as the UK government had to launch its largest-ever peacetime repatriation effort to bring its citizens back home at the expense of the British taxpayer.

Additionally, 22,000 employees sadly lost their jobs due to the collapse, and thousands more faced an uncertain future. Meanwhile, the company's creditors, including hotels and suppliers, were left with unpaid bills, leading to financial difficulties for many businesses (Whyte,2019).

The company's mounting debts of £1.7 billion proved too much to bear (Sims, 2019). Despite its efforts to secure a £1.1 billion rescue package, it ultimately failed. The collapse of this travel giant had a significant impact on the travel industry and resulted in thousands of employees losing their jobs.

Despite its impressive 178-year history, Thomas Cook struggled to keep pace with changing trends in the travel industry. It was once renowned as a top provider of package holidays, utilising its global reputation and financial power to forge relationships with many small-scale tour operators, hotels, and caterers. However, in recent years, it has lagged behind the competition.

The reasons behind Thomas Cook's failure were multifaceted. The rise of online travel booking platforms such as Expedia and Booking.com had led to a decline in the company's profits. Additionally, the company's outdated business model, high operating costs, and failure to adapt to changing consumer preferences played a crucial role in its downfall.

The company's massive debt made it difficult for Thomas Cook to secure additional funding, ultimately leading to its demise.

By examining the underlying factors that contributed to Thomas Cook's downfall, other tour operators can gain valuable insights that can help them avoid repeating the same mistakes (McCormick, 2019).

Reasons for Thomas Cook's failure

Financial Decline

Between 2013 and 2018, Thomas Cook experienced a significant increase in its net debt, which had a direct impact on its earnings per share. This decline in earnings can be observed in Figure 1, which presents a comparison of the company's Basic Earnings Per Share (EPS) and underlying EPS over this period in conjunction with the evolution of its net debt.

It is worth noting that Basic EPS represents the company's earnings per share after deducting all the expenses and taxes, while underlying EPS excludes exceptional items such as restructuring costs or gains/losses on the disposal of assets.

These two metrics are commonly used to evaluate a company's profitability and financial health. The information mentioned here is available in the Thomas Cook Group annual report of 2018, which provides a detailed analysis of the company's financial performance during the stated periods.

Figure 1. Basic EPS, Underlying EPS and Net Debt between 2013-2018

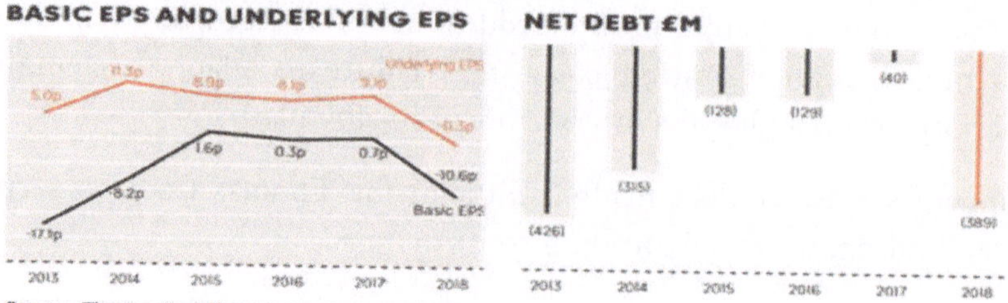

Source: Thomas Cook Group annual report, 2018

On September 9th, 2019, Thomas Cook faced a significant crisis that sent shock waves throughout the travel industry. Its lenders, which included the Royal Bank of Scotland and the Lloyds Banking Group, demanded that it secure a seasonal standby facility of GBP 200 million in addition to the GBP 900 million it had already received as fresh capital.

This extra funding was meant to help the company sustain its operations during the lean winter period. However, Thomas Cook struggled with financial problems for some time and could not raise GBP 200 million (Stubley,2019; Sims,2019). As a result, the company had no option but to cease trading, leading to a devastating collapse.

The insolvency of Thomas Cook, which had been in operation for 178 years, was the largest ever in the travel industry and a huge blow to its employees and their families. The collapse of Thomas Cook had far-reaching consequences.

Outdated Business Model

In 2010, Thomas Cook initiated a substantial expansion plan to increase the number of its physical stores. As part of this strategy, it joined forces with Co-operative Travel to create the most prominent travel agent on the high street. However, the collaboration resulted in a catastrophic outcome for the company.

Thomas Cook's troubles worsened in 2016 when it spent an astonishing £82 million to acquire complete control of Co-op's travel arm. Even this move could not reverse the damage caused by the partnership, and in 2017, Thomas Cook had to concede that the collaboration was a massive failure (Butler,2016).

Meanwhile, the company was also grappling with its mounting debts, which had skyrocketed to a staggering £804 million in 2010. To address this issue, Thomas Cook adopted a cost-cutting strategy that included laying off employees and seeking additional funding from lenders. The company was able to secure emergency funding of £100 million to keep its operations moving forward. However, the challenges for Thomas Cook did not end there (Butler,2016).

With the advent of the Internet and the rise of low-cost airlines, consumers began shifting to online platforms to book their holidays. This change in consumer behaviour had a significant impact on Thomas Cook's business model, which heavily relied on its physical stores.

In addition, the company faced intense competition from smaller, online-only travel businesses that enjoyed lower overheads due to their lack of physical stores. These factors further compounded the company's problems and made it difficult to stay competitive in the market (Thomas Cook Group annual report of 2018).

Merger with MyTravel

Back in 2007, Thomas Cook and UK-listed rival MyTravel, announced their decision to merge in a £2.7bn deal. The merger was expected to create a global tour operator powerhouse across the UK, Germany, France, and Canada. The new company would have up to 1,100 travel shops on British high streets alone and become a significant player in the industry. However, the merger turned out to be a catastrophic mistake in several ways (Spikes et al.2007).

Firstly, it resulted in a £74m windfall for the bankers and lawyers involved in the deal, which added to the company's expenses. Secondly, the merger loaded the new company with high debts, which eventually brought it down.

Despite high expectations of profitability, it soon became apparent that profits were far lower than expected in the first year of the merger. It must be noted, however, that MyTravel reported its first annual profit since 2001.

In the aftermath of the merger, a major scandal rocked the company, leading to the resignation of Thomas Middelhoff, the chairman of the Thomas Cook supervisory board, who had played a key role in bringing the two companies together.

The scandal revealed that Middelhoff had been involved in the Oppenheim-Esch Fund (Oppenheim was the German Private bank that had a real estate fund with Josef Esch that Middelhoff financed), which had come under scrutiny from the public prosecutor's office. This led to a preliminary investigation into Middelhoff's involvement, adding to the growing controversy surrounding his role at Thomas Cook and Arcandor.

The fallout from the scandal was significant, with Arcandor being forced to seek government guarantees and loans in May 2009. The scandal also raised serious questions about the oversight and governance of Thomas Cook and led to widespread calls for greater transparency and accountability in the company's operations. These developments were early warning signs of the merged business's troubles.

Fast-forward to May 2018, the company revealed a debt pile of £1.2bn, which was a significant burden on its finances. Furthermore, the company had to record a £1.1bn write-off in its balance sheet from its ill-fated acquisition of MyTravel, which further added to its financial woes (Thomas Cook Annual Report, 2018).

Thomas Cook's senior team obscene remuneration packages at the expense of a failing business

The failure of Thomas Cook has sparked a renewed discussion about executive remuneration. The company's senior management team has faced harsh criticism for the generous bonuses they have received over the years, despite the company's financial struggles. The Financial Times has reported that the Chief Executives at Thomas Cook were paid a total of £18.7m over the past decade, with the last CEO earning over £8m in just five years (McCormick,2019).

This has led to calls for company remuneration committees to adopt a more sustainable system that is more closely aligned with performance. It is essential that senior executives pay packages reflect the risk of failure and that rewards are given only when performance is achieved (Farmborough, 2019).

The UK government is pushing for the Insolvency Service to expedite its investigation into the situation and determine the involvement of the senior team. The collapse of Thomas Cook highlights the importance of aligning executive pay with performance and taking the necessary steps to ensure that failure is not rewarded.

It is essential to ensure that senior executives are held accountable for their actions and that their remuneration is tied to their performance in the company. Moreover, it is necessary to consider whether director bonuses should be revoked if they act against the company's interests. This approach will encourage executives to make decisions that prioritise the long-term success of the company over their personal gain.

Self-Interest versus Company's Interest

Oliver Richardson, a representative of Trade Union Unite, has accused the directors, lawyers, accountants, and auditors who advised Thomas Cook of being responsible for the company's collapse. According to Richardson, these professionals prioritised their financial gain over the long-term interests of the company, which ultimately led to its downfall. This accusation highlights the issue of professionals putting their own interests ahead of those of the companies they are supposed to serve (Irish Examiner, 2019).

The collapse of Thomas Cook led to hedge fund managers making a profit of $250 million through Credit Default Swaps (CDS) bets on the company's outstanding debts.

This has raised further criticism about the way the company handled its accounts, particularly after Ernst and Young (EY) was appointed as the new auditor in 2017, replacing Price Water House Coopers (PWC), which had been signing off the organisation's books for years in exchange for a staggering £4 million in annual fees (Thomas Cook Group's Annual Report, 2018).

This highlights the need for companies to have better financial management practices and for auditors to be more vigilant in their oversight (Thomas Cook Group Annual report, 2018).

Conclusion

The collapse of Thomas Cook has been a complex and multi-faceted story, with a number of external factors playing a role in the company's downfall.

However, it is important to note that many of the issues that led to the company's collapse were also self-inflicted.

For example, while it is true that Thomas Cook was impacted by external factors such as political instability, natural disasters, and terrorism, the company's management also allowed its debts to grow exponentially while failing to adapt to the changing landscape of the travel industry. This left the company in a precarious financial position that ultimately led to its collapse.

Despite this, the collapse of Thomas Cook resulted in criticism being aimed at a range of stakeholders, including bondholders, banks, the UK government, and Thomas Cook's largest shareholder, Fosun. Many have argued that these stakeholders could have done more to save the company, particularly given the significant impact that the collapse has had on thousands of employees and customers.

One of the key issues that has been highlighted is the way in which Thomas Cook raised funds in 2013. At that time, the company raised £425m from shareholders, which initially appeared to be a positive development. However, the 2013 bailout was funded by 'only' £425m of equity, with the balance of over £1bn being secured through debt. This decision left the company with a substantial debt burden that it was unable to service in the years that followed.

By May 2018, Thomas Cook was worth £2bn, and at this point, shareholders and the senior team should have reviewed their books and taken steps to clear the company's debts. However, no decisive action was taken, and the company's debt continued to grow, leaving it increasingly vulnerable to financial instability.

Tighter controls and regulations from Governments are needed to prevent another failure of this magnitude from happening again in the future. It is also important to ensure that stakeholders at all levels take responsibility for their part in the company's collapse and work to prevent similar scenarios from occurring in the future.

The failure of Thomas Cook also highlights the challenges facing the travel industry, which is vulnerable to changing consumer preferences, geopolitical risks, and economic uncertainty.

Thomas Cook's 178-year history teaches us that businesses need to stay relevant and adaptable in an ever-evolving market. The company's ultimate failure serves as a cautionary tale for other businesses, reminding them that they must be nimble and stay ahead of the curve to survive in a fast-paced, technology-driven world.

CHAPTER EIGHT

The dispute between Indian Low Cost Carrier Go First and engine manufacturer Pratt & Whitney

Abstract

The Indian aviation industry has experienced a rapid expansion in recent years. However, it is unfortunate that several of the leading Indian airlines have been unable to translate this growth into financial success.

The latest setback for the industry is the insolvency plea filed by Go First, a prominent Indian low-cost carrier (LCC), before the Indian National Company Law Tribunal (NCLT).

Go First has been struggling to cope with engine issues in its Airbus A320neo aircraft, which were supplied by Pratt & Whitney's International Aero Engines, LLC. The airline blames the engine supplier for these issues.

Due to the increasing number of failing engines, the airline was forced to ground 25 of its planes, almost half of its entire fleet. Nonetheless, reports suggest that the airline was already in debt of nearly $798 million and was only using the engine issue as a reason to ground its planes

This case study highlights the operational challenges faced by Go First in a domestic Indian market dominated by LCC's.

Key Words: Bankruptcy, Growth, Low-Cost Carriers, Voluntary Insolvency.

Background

Go First, formerly known as GoAir, is a low-cost airline founded in 2005 by Jehangir Wadia, the son of Indian industrialist Nusli Wadia. The airline is headquartered in Mumbai, Maharashtra, and began operating on November 4, 2005.

It operates domestic flights to 34 destinations within India and international flights to 8 destinations across Asia, and the Middle East.

The airline had planned to increase its fleet size to 100 aircraft by 2025 and had announced its intention to launch an Initial Public Offering (IPO) to raise Indian rupees ₹36 billion (US$450 million). However, the airline encountered issues with its fleet of Airbus A320neo aircraft powered by Pratt & Whitney's geared turbofan (GTF) engines (Thomas et al.2023).

In 2023, the airline faced critical operational issues that led to the cancellation of multiple flights (Go First,2023). In a recent press release, the airline made it clear that it had no choice but to take this step as the engine issues posed a serious threat to the safety and reliability of its flights.

In May 2023, Go First, declared bankruptcy due to a confluence of factors that ultimately led to significant financial losses. The primary reasons cited for the airline's financial distress included the ongoing Covid-19 pandemic, which resulted in reduced demand for air travel and led to the cancellation of many flights.

Additionally, the airline faced increased fuel costs, which made it difficult to maintain profitability. Lastly, engine procurement issues also contributed to the airline's financial woes. Despite its best efforts to navigate these challenges, Go First ultimately succumbed to the mounting pressures and was forced to declare bankruptcy.

As a result of this bankruptcy, many lessors have terminated their contracts with Go First. However, the airline has been protected from repossession by a court injunction, which has made it difficult for lessors to recover their aircraft.

When did the problem start?

In 2016, the airline received its first A320 neo aircraft powered by GTF engines after ordering 72 narrow-body jets from Airbus. Pratt & Whitney is the exclusive engine supplier for the airline's A320neo aircraft fleet, which makes up around 90% of Go First's fleet.

Despite choosing the engines because the United States (US) firm offered better fleet-management terms and the engines would be more fuel-efficient, quieter, and require less service, the GTF engines were found to have issues on the fan blades, an oil seal, and the combustion chamber lining (Shukla, 2023).

The engine issue was first noticed in 2017, a year after Go First received its initial GTF-powered A320 neo aircraft. It's worth noting that Go First was one of many customers encountering engine problems, as India's budget airline IndiGo had reported similar issues.

Since 2020, Go First aeroplanes have been unable to fly due to an acute shortage of spare parts and an unfortunate delay in the delivery of retrofitted engines from Pratt and Whitney, the engine manufacturer (Reuters, 2023).As a result, almost half of their fleet has been grounded, causing severe disruptions to their flight schedules and frequent cancellations.

Pratt and Whitney were ordered by an arbitration court order in Singapore to provide 'at least ten serviceable spare leased engines by 27 April 2023' to Go First. However, the engine manufacturer failed to comply and rejected Go First's allegations, which it found unacceptable.

This has caused a significant setback for Go First, which is struggling to cope with the financial losses and reputation damage caused by the cancellations and delays (Reuters, 2023).

The Legal Wrangling

The global aviation industry is a complex and multifaceted sector, with aircraft travel being a crucial component that involves crossing borders. Given the unique nature of this mode of transportation, it is essential to have a robust international framework that clearly defines the rights of lessors and operators in the event of bankruptcy and liquidation.

To this end, the Convention on International Interests in Mobile Equipment (2001) ('Cape Town Convention') and its Protocol to the Convention in Mobile Equipment on matters specific to Aircraft Equipment (2001) ('Aircraft Protocol') have been instrumental in providing creditors with various remedies for airline defaults.

These treaties are particularly crucial for the aviation industry's stability, as highlighted by Honnebier (2005). However, despite joining the Cape Town

Convention in 2008, India still needs to legislate this treaty legally. The primary reason is that Indian local laws prioritise bankruptcy protection over repossession requests, leading to ambiguity and non-compliance.

This situation is unacceptable and requires immediate attention to ensure India complies with its international obligations, as pointed out by Variath and Dutta (2023).

For several years, the Indian government has failed to modify domestic law to implement the provisions of an international convention, leading to a negative outlook from the global aviation watchdog's Aviation Working Group (AWG) for the Indian Aviation Industry. This non-compliance can have severe consequences for the financing and leasing of Indian airlines.

However, India took a significant step towards compliance with the Cape Town Convention on October 3, 2023, when the Indian Ministry of Corporate Affairs issued a notification to amend Section 14(3)(a) of the Insolvency and Bankruptcy Code 2016.

The notification clarified that the provisions shall not apply to transactions under the Cape Town Convention relating to aircraft, aircraft engines, airframes and helicopters. This announcement was met with a favourable response from the aircraft leasing community, as it confirmed that the Cape Town Convention would take precedence in future insolvency hearings involving aircraft assets. The AWG issued a positive notice in response to the ruling.

Conclusion

Go First was a well-known Indian airline with a stellar reputation for its excellent performance and successful expansion strategy until 2020. Unfortunately, the company faced multiple engine failures that caused a significant cash flow problem that year.

Despite its efforts to recover from the setback, Go First's financial struggles persisted, and, on May 10th, 2023, the Indian National Company Law Tribunal granted the airline bankruptcy protection.

The collapse of Go First has been a significant blow to the Indian airline industry. It has highlighted the intense competition in the sector, which has seen one of the most robust air traffic recoveries during the pandemic.

In response to the situation, SpiceJet, a popular Indian low-cost carrier, has proposed a plan to reorganise Go First and merge it with its operations.

This proposal has generated much interest, as it could potentially revitalise the struggling airline and create a more formidable competitor in the Indian airline market. However, it remains to be seen whether the proposal will be accepted and how it will impact the industry.

This is an evolving story that we will be keeping a close eye on.

CHAPTER NINE

Conclusion

The aviation industry has faced significant challenges in recent years, with the sector reaching its peak in 2012. Since then, many airlines have been forced to shut down, resulting in a considerable loss of jobs and financial damage to the industry. However, nothing could have prepared the aviation industry for the unprecedented impact of the COVID-19 pandemic that hit the world in 2020.

The pandemic brought a wave of devastation to the aviation industry, with hundreds of aircraft grounded and airports operating at minimal capacity. As a result, passenger revenues have been decimated, and the industry has been hit hard. The pandemic has particularly affected smaller airlinesstruggling to resume profitable operations due to a lack of government support and access to affordable capital. In some cases, these airlines have ceased trading altogether.

The immediate effects of the pandemic were temporary, but the interaction of the pandemic with other strategic factors, such as globalisation and the digital economy, will undoubtedly have long-term consequences. The aftermath of COVID-19 further reveals the inherent vulnerability of the aviation industry, which is not a fleeting issue but a formidable systemic challenge that requires prompt and decisive actions.

It is essential to note that COVID-19 was undoubtedly one of the reasons for some airline failures, but most failed airlines had financial problems that pre-date the pandemic outbreak and were already well documented.

Therefore, COVID-19 is likely to be seen as the scapegoat for failed airlines, and it would be shortsighted to attribute reaching the situation entirely to it. The aviation industry was already struggling before the pandemic, and the pandemic only exacerbated the existing issues. The industry needs to address its underlying problems and undertake significant structural reforms to ensure its long-term sustainability.

The aviation industry is a highly competitive and ever-changing sector that encompasses multiple factors determining airlines' triumph or downfall. Airlines don't fail for a single reason but rather a combination of factors, including poor financial management, an inability to adapt to changing market conditions, lack of innovation, and bad management decisions.

These factors are common among airlines that had already failed before the pandemic, and they serve as a reminder that proactive and innovative business approaches are essential to achieving long-term success in the aviation industry.

The case studies in this book showcase the hurried and imitative behaviour of certain airlines that failed to keep up with the industry's dynamic advancements. These airlines needed help to recognise the significance of innovation and strategic planning in the aviation industry, which resulted in their eventual downfall.

One of the most recent airline bankruptcy cases is Go First in India, which is the last case study in the book (Chapter Eight). The case study highlights how technological issues can have wide-reaching consequences for other airlines. Go First experienced engine issues on its Airbus 320-NEO aircraft, and the case between Go First and Pratt and Whitney is still pending in the courts at the time of writing.

When the scale of the problem was revealed, some experts predicted that more airlines could follow suit and take Pratt and Whitney to court to claim compensation. However, since then, Pratt and Whitney has put its fleet management plan in place and worked to compensate affected companies, and there is more sense that the issue will be resolved with too few airline casualties.

Nonetheless, despite the setbacks caused by various issues, airlines have risen to the challenge and entered a new decade with determination and resilience to overcome this extraordinary obstacle facing the world's aviation industry.

The emergence of new airlines with clean balance sheets over the past year highlights the industry's resilience and ability to adapt to changing circumstances. This positive development enables airlines to fortify their financial positions and secure their future despite uncertainties. However, airlines must exercise vigilant management and strategic planning to avoid scenarios like airline failures and bankruptcies.

The aviation industry is an intricate and demanding sector requiring companies to be agile and adaptable to survive. Airlines that keep up with the ever-changing industry trends and seize opportunities are the ones that achieve long-term success.

Each case study in this book reminds us that airlines that fail to innovate and adapt to the industry's changing landscape are bound to suffer the consequences.

References

1. Airlines: Riding the Recovery. Aviation Leaders Report 2023 [Online]. Available from:*https://kpmg.com/ie/en/home/insights/2023/01/aviation-leaders-report-2023/airlines-riding-the-recovery.html* [Accessed 30 January 2023].

2. Buckley, J. (2023) How the pandemic killed off 64 airlines [Online]. Available on: *www.edition.cnn.com/travel/article/pandemic-airline-bankruptcies/index.html* [Accessed 1 March 2023].

3. IATA (2023) Air Passenger Market Analysis [Online]. Available from: *https://www.iata.org/en/iata-repository/publications/economic-reports/air-passenger-market-analysis-december-2023/#:~:text=Asia%20Pacific%20airlines%20more%20than,%25%20under%20pre%2Dpandemic%20levels* [Accessed 10 January 2024].

4. Airlines: Riding the Recovery. Aviation Leaders Report 2023 [Online]. Available from:*https://kpmg.com/ie/en/home/insights/2023/01/aviation-leaders-report-2023/airlines-riding-the-recovery.html* [Accessed 30 January 2023].

5. Bellovary, J.L., Giacomino,D.E. and Akers,M.D. (2007) A Review of Bankruptcy Predication Studies: 1930 to Present. *Journal of Financial Education* (33), p.1-42.

6. Borenstein, A. and Rose, N.L. (2017) Do Airlines in Chapter 11 Harm Their Rivals?: Bankruptcy and Pricing Behavior in U.S. Airline Markets. 1st ed. Andesite Press.

7. Cambier Y., Rubin S., Barbany M., Bachan A. COVID-19: Fleet Outlook and Impact on Lessors and MROs (2020) [Online]. Available on: *https://www.icf.com/insights/transportation/covid-19-fleet-oulook-impact-lessors-mro [Accessed on: 1 August 2023].*

8. Dunn, G. (2017) `Analysis: Airline start-ups and failures in 2017. F*light Global*, December 27.

9. Iverson, B. (2018) Get in line. Chapter 11 restructuring in crowded bankruptcy courts. Management Science, 64 (11), p.5370-5394.

10. Reger, R.K. (1990) Managerial thought structures and competitive positioning. Mapping strategic thought. Chichester: John Wiley and Sons.

11. Tisdall, L., Zhang, Y., Zhang, A., (2021). Covid-19 impacts on general aviation – comparative experiences, govern- mental responses and policy imperatives. *Transport Policy*, P.110, 273–280.

12. Yamaguchi, K. (2022) Productivity impact of government-led bailout of Japan Airlines. *Journal of Asian Transport Studies* [Online], 8 (4),26 February 2022. Available from: *https://www.sciencedirect.com/science/article/pii/S2185556022000010* [Accessed 1 May 2023].

13. Blue Air (2023) [Online]. Available from: h*ttps://en.wikipedia.org/wiki/Blue_Air* [Accessed: 23 November 2023].

14. Dunn, G. (2023) Europe probes Romanian state aid for grounded Blue Air [Online]. Available from: *https://www.flightglobal.com/airlines/europe-probes-romanian-state-aid-for-grounded-blue-air/152872.article* [Accessed 20 August 2023].

15. European Commission (2024) Commission orders recovery of €33.84 million in incompatible state aid from Blue Air [Online].Available from:*https://ec.europa.eu/commission/presscorner/detail/ne/ip_24_848* [Accessed 18 February 2024].

16. European Commission (2023) State Aid: Commission opens in-depth investigation into Romanian Support measures in favour of Blue Air [Online]. Available from: *https://ec.europa.eu/commission/presscorner/detail/en/IP_23_1364* [Accessed 20 October 2023].

17. Journal, R. (2023) Blue Air Became Insolvent [Online]. Available from: *https://www.romaniajournal.ro/business/blue-air-became-insolvent/* [Accessed 23 November 2023].

18. Reports of Cases Judgement of the General Court (Tenth Chamber, Extended Composition) (2023) [Online]. Available from: *https://op.europa.eu/en/publication-detail/-/publication/6ee89034-092c-11ee-b12e-01aa75ed71a1/language-en* [Accessed 23 September 2023].

19. Airports Commission (2015) Economy: Wider Economic Impacts Assessment. July 2015.

20. Annual Revenue of Flybe group plc (2009 to 2018) [Online]. Available from: *https://www.statista.com/statistics/1007898/revenue-flybe-plc/* [Accessed: 20 April 2020].

21. Burghouwt, G. and Redondi, R. (2013) Connectivity in Air Transport Networks. An Assessment of Models and Applications. *Journal of Transport Economics and Policy, 47 (1), p.35-53.*

22. Flottau, J.(2020) Crisis mode : COVID-19 turns industry challenges into drama; IATA predicts up to $113 billion in revenue losses; Flybe collapses and more consolidation is expected [Online]. Available from: *https://trid.trb.org/view/1700891* [Accessed 01 May 2020].

23. Flybe Annual Report (2017-18) [Online]. Available from: *https://flybeplc.github.io/deal/rns/Flybe_Annual_Report_2017-18.pdf* [Accessed 20 April 2023].

24. Flybe (1979-2020) [Online]. Available *from:https://centreforaviation.com/data/profiles/airlines/flybe-1979-2020-be* [Accessed: 23 April 2020].

25. Gossling, S., Scott, D. and Hall, C.M. (2021)Pandemics, tourism and global change: a rapid assessment of COVID-19.*Journal of Sustainable Tourism*, 29 (1).

26. Grant, J. (2020) Flybe – Can Regional be too Regional [Online].Available from*: https://www.oag.com/blog/flybe-can-regional-be-too-regional* [Accessed 18 September 2020]

27. Hockley, T. (2020) Flybe rescue: why government maybe putting the green revolution at risk. British Politics at LSE.

28. Laird, J. and Mackie, P. (2018) Wider Economic Impacts of Regional Air Connectivity. Peaks Economics.

29. Lawrie, E. (2020) Flybe collapse: Five things that went wrong [Online]. Available from*: https://www.bbc.co.uk/news/business-51749882* [Accessed 23 April 2020].

30. Rose, N.L. (2014) Economic Regulation and its Reform: What we have learned.1st edition. University of Chicago Press.

31. UK regional airlines: Flybe or not Flybe? Connectivity! (2020) [Online]. Available from: *https://centreforaviation.com/analysis/reports/uk-regional-airlines-flybe-or-not-flybe-connectivity-509290* [Accessed 30 April 2020].

32. Rabinovici, I. (2019) The Application of EU Competition Rules in the Transport Sector in 2019. *Journal of European Competition Law and Practic*e, 11 (5-6), p.327-332.

33. Rowland, B. (2020) Lessons from Flybe-European Regional Airline Networks and Hub Access [Online]. Available from: *https://www.oag.com/blog/lessons-from-flybe-european-regional-airline-networks-and-hub-access.* [Accessed 30 April 2020].

34. Topham, G. (2020) Flybe: airline collapses two months after government announces rescue [Online]. Available from: *https://www.theguardian.com/business/2020/mar/05/flybe-collapses-two-months-after-government-announces-rescue* [Accessed 10 April 2020].

35. Topham, G. (2020) Flybe saved after ministers and investors seal rescue deal [Online]. Available from: *https://www.theguardian.com/business/2020/jan/14/flybe-saved-after-successful-last-ditch-talks#:~:text=The%20package%20of%20measures%20includes,flights%20before%20the%20March%20budget* [Accessed 10 April 2020].

36. Seetram, N., Song, H. and Page, S.J. (2013) Air Passenger Duty and Outbound Tourism Demand from the United Kingdom. *Journal of Travel Research,* 53 (4).

37. Calder, S. (2022) Flybe returns with £20 flight from Birmingham to Belfast [Online]. Available from: *https://www.independent.co.uk/travel/news-and-advice/flybe-flight-birmingham-belfast-uk-b2057073.html* [Accessed 14 April 2022)

38. Cook, G.N. and Bullig, B.G. (2017) Airline Operations and Management: A Management Textbook. 1st edition.Routledge.Taylor and Francis Group.

39. Crans, B.J.H and Nath, R. (2010). Aircraft Repossession and Enforcement: Practical Aspects Volume II (International Bar Association Series): 2 (Kluwer Law International: International Bar Association).1st edition. Wolters Kluwer. Law and Business.

40. Dey, M. and Davey, J. (2023) UK regional airline Flybe ceases trading, and cancels all flights [Online]. Available from: *https://www.reuters.com/business/aerospace-defense/uk-airline-flybe-put-administration-cancels-scheduled-flights-2023-01-28/* [Accessed 23 February 2023]

41. Flybe Limited (2023) [Online]. Available from: *https://find-and-update.company-information.service.gov.uk/company/12875147* [Accessed: 02 February 2023].

42. Joint Administrators' proposals. Flybe Limited -in Administration (2023) [Online]. Available from: *https://www.headforpoints.com/wp-*

content/uploads/2023/03/Flybe-restructuring.pdf [Accessed 21 February 2023].

43. Joint Administrators progress report for the period 28 January 2023 to 27 July 2023: Flybe Limited -in administration [Online]. Available from: *https://find-and-update.company-information.service.gov.uk/company/12875147/filing-history* [Accessed 01 September 2023].

44. Mayling, S. (2022) Flybe announces routes as tickets go on sale [Online]. Available from: *https://travelweekly.co.uk/news/air/flybe-announces-routes-as-tickets-go-on-sale* [Accessed 24 April 2023].

45. Perry, D. (2023) Flybe administrators detail likely creditor payouts [Online]. Available from: *https://www.flightglobal.com/airlines/flybe-administrators-detail-likely-creditor-payouts/154723.article*[Accessed 24 September 2023].

46. UK Civil Aviation Airline Licensing Decisions. Decisions On Air Transport Licences, Route Licenses and Scarce Capacity (2021) [Online].Available from: *https://www.caa.co.uk/commercial-industry/airlines/licensing/requirements-and-guidance/airline-licensing-decisions/* [Accessed 21 February 2023].

47. Arora.D. and Ravi.S. (2019) Issues and Challenges of Indian Aviation Industry: A case of Jet Airways. *Review of Management*, 9 (3-4), p.1557-1562.

48. Daddikar, P.V. and Shaikh, A.R. (2014). Impact of Mergers and Acquisitions on Surviving Firm's Financial Performance: A Study of Jet Airways Ltd. Pacific Business Review International, 6 (8), p45-51.

49. Dalal S. (2006). *How did Jet go so off-figure on Sahara? [Online]. Available From: http://www.suchetadalal.com/?id=f2dbe382-85c2-9361-492e8c1f69a5&base=sub_sections_content&f&t=How+did+Jet+go+so+off-figure+on+Sahara%3F* [Accessed 1 March 2022]

50. Doganis, R. (2001). The airline business in the 21st century. 1st ed.Routledge.

51. Ghosh K. (2018). *Crisis-hit Jet Airways is still dealing with the baggage from the Air Sahara deal [Online]*. Available From: https://scroll.in/article/897458/crisis-hit-jet-airways-is-still-dealing-with-the-baggage-from-the-air-sahara-deal [Accessed 20 February 2022].

52. Indian Aviation Industry (2021) [Online]. Available from: *https://www.ibef.org/industry/indian-aviation* [Accessed 3 February 2022].

53. Jet Airways (India) [Online]. Available from: *www.statista.com/companies/c/25128977/jet-airways--india* [Accessed 1 November 2021].

54. Kathpal, S. and Akhtar, A. (2021), "Fall of a Titan: understanding the Jet Airways crisis. *Journal of the case association,* 17 (4), p. 569-587.

55. Krishnan, R.T.(2008) The Indian Airline Industry in 2008. Indian Institute of Management Bangalore.

56. Mahalakshmi N. (2013). *Jet, Sahara, Go.* [Online]. Available from: https://www.business-standard.com/article/markets/jet-sahara-go-107041601035_1.html [Accessed 1 March 2022].

57. Public liquidation report (PLR) 3 Jet Airways (ex-article 73A of the Dutch Bankruptcy Act) (2019) [Online]. Available from: https://www.potjonker.nl/app/uploads/2020/08/Jet-Airways-India-Ltd.-liquidation-report-nr.-3-English-translation.pdf [Accessed 3 February 2021].

58. O,Connell, J.F. and Willams, G. (2006). Transformation of India's Domestic Airlines: A case study of Indian Airlines, Jet Airways, Air Sahara and Air Deccan. *Journal of Air Transport Management* , 12 (6), p.358.374.

59. Saran R. and Mehra P. (2012) [Online). *Acquisition of Air Sahara by Jet Airways: Domination of Naresh Goyal over Indian skies.* Available from: *https://www.indiatoday.in/magazine/cover-story/story/20060206-jet-airways-acquisition-of-air-sahara-domination-of-naresh-goyal-over-indian-skies-786030-2006-02-06* [Accessed 3 February 2022].

60. Shastri, M. (2014) Indian Aviation: Flying Through Turbulence. International *Journal of Engineering Research and Management Technology,* 1 (6), p127-128.

61. Statista: Jet Airways (2019) [Online]. Available from:*https://www.jetairways.com/* [Accessed 1 November 2021]

62. Sun, S. (2021) Key financial figures of Jet Airways FY 2014-2019 [Online]. Available from: *www.statista.com/statistics/587559/financial-performance-jet-airways/* [Accessed 23 February 2022].

63. The Economic Times (2023) How Jet Airways founder Naresh Goyal built a successful airline and then sank it [Online].Available from:

https://economictimes.indiatimes.com/industry/transportation/airlines-/-aviation/how-jet-airways-founder-naresh-goyal-built-a-successful-airline-and-then-sank-it/articleshow/103313477.cms?track404=1&from=mdr [Accessed 23 August 2023].

64. Butler, S. (2016) Co-operative Group secures £82m windfall in sale of travel division [Online]. Available from:*https://www.theguardian.com/business/2016/dec/06/co-operative-group-sale-thomas-cook-travel-agencies* [Accessed 10 August 2019].

65. Economic Investigations. (2015) Handout: Barriers to Entry - Tour Operators [Online]. Available from: *http://peped.org/economicinvestigations/handout-barriers-to-entrytour-operators* [Accessed 4 January 2019].

66. Farmborough, H. (2019) Thomas Cook Pay Row Highlights Bigger Issue [Online].Available from: *https://www.forbes.com/sites/heatherfarmbrough/2019/09/26/thomas-cook-pay-row-highlights-bigger-issue/?sh=465efb93968a* [Accessed 20 December 2019].

67. Irish Examiner (2019) Thomas Cook staff prepare to be out of pocket on first payday since collapse [Online]. Available from: *https://www.irishexaminer.com/business/arid-30953748.html* [Accessed 01 December 2019].

68. Leadbeater, C. (2019) The history of Thomas Cook, from tours for teetotallers to boozy packages in Spain. The Telegraph [online]. Available from: *https://www.telegraph.co.uk/travel/tours/history-of-thomas-cook/* [Accessed 4 January 2019].

69. McCormick, M. (2019) Thomas Cook: Collapse of UK's oldest travel group reverberates around the world - as it happened. *Financial Times* [Online]. Available from https://www.ft.com/content/18c6356f-d806-3fef-9ff7-29fb80a343c7 [Accessed 01 September 2019].

70. Middleton, P. (2011) Thomas Cook bonuses fuel investors anger [Online]. Available from:*https://www.ft.com/content/d573c284-361b-11e0-9b3b-00144feabdc0* [Accessed 01 September 2019].

71. Schwartz, K., Tapper, R. and Font, X. (2009). A Sustainable Supply Chain Management Framework for Tour Operators. Journal of Sustainable Tourism, 16 (3), P.298-314.

72. Sims, S. (2019) How Could Travel Giant Thomas Cook Fail? The New York Times [Online]. Available from:

https://www.nytimes.com/2019/09/23/travel/why-thomascook-travel-collapsed.html [Accessed 4 January 2019].

73. Stubley, P. (2019) Thomas Cook blames Brexit for £1.5bn loss. Independent [Online]. Available from *https://www.independent.co.uk/news/business/thomas-cook-shareprice-profit-loss-results-brexit-travel-flights-a8916061.html* [Accessed 24 February 2019].

74. Spikes, S. and Blitz, R. (2007) MyTravel agrees merger with Thomas Cook [Online]. Available from:*https://www.ft.com/content/a5c4380a-ba34-11db-89c8-0000779e2340* [Accessed 24 February 2019].

75. Thomas Cook Annual Report (2018). Thomas Cook Group plc [Online]. Available from: *https://www.annualreports.com/HostedData/AnnualReports/PDF/LSE_TCG_2018.pdf* [Accessed 23 February 2019].

76. Thomas Cook (2019) Share Price [Online]. Available from *https://www.thomascookgroup.com/investors/share_price* [Accessed 23 February 2019].

77. Whyte, P. (2019) Debt, Egos and Bad Decisions: How Thomas Cook Failed to Adapt to a New Era of Travel [Online].Available from https://skift.com/2019/09/30/debt-egos-and-bad-decisions-how-thomas-cook-failed-to-adapt-to-a-new-era-of-travel/ [Accessed 20 September 2019].

78. Go First (2023) [Online]. Available from: *www.en.wikipedia.org/wiki/Go_First* [Accessed 10 August 2023].

79. Honnebier, P. (2005) The Cape Town Convention and the Aircraft Equipment Protocol: Protecting the Registered secured Interests of Airline Lessees. *Air and Space Law*, 30 (1), p.27-35.

80. Reuters (2023) The engine dispute at the heart of Go First's bankruptcy filing [Online]. Available from: *www.reuters.com/business/aerospace-defense/engine-dispute-heart-go-firsts-bankruptcy-filing-2023-05-03/* [Accessed 30 August 2024].

81. Shukla, A. (2023) Go First: What went wrong with Indian airline? [Online]. Available from: *www.bbc.co.uk/news/world-asia-india-65458136* [Accessed 23 August 2023].

82. Thomas, C., Skariachan,D. and Mehta, T.(2023) India's Go First files for bankruptcy, blames Pratt and Whitney engines [Online]. Available from: *www.reuters.com/business/aerospace-defense/indias-cash-strapped-go-first-airways-suspends-flights-et-2023-05-02/* [Accessed 23 August 2023].

83. Variath, A.A. and Dutta, S.(2023). India and Cape Town Convention: Contextualizing Aviation Industry Bankruptcy in the Case of Go First. *Dharmashastra Nat'l L. Univ. L. Rev.*, p.15.

9 789362 610706

Printed at Repro India Ltd.